FROM THESE LANDS

FROM THESE LANDS

Sharing Our Natural and Cultural Heritage

FEATURING OBJECTS FROM EVERY STATE AND INHABITED TERRITORY

NATIONAL MUSEUM OF NATURAL HISTORY

SMITHSONIAN BOOKS
WASHINGTON, DC

Published by Smithsonian Books
PO Box 37012, MRC 513
Washington, DC 20013
smithsonianbooks.com

DIRECTOR: Carolyn Gleason
SENIOR EDITOR: Jaime Schwender
PRODUCTION EDITOR: Julie Huggins
DIGITAL IMAGING TECHNICIAN: Bill Whitcher

EDITED BY Jamie Greene
DESIGNED BY Kevin Coochwytewa

National Museum of Natural History
SANT DIRECTOR: Kirk Johnson
WRITERS: Angela Roberts Reeder and Juliana Olsson
PUBLICATIONS TEAM: Junko Chinen, Jill Johnson, Michael Lawrence, Juliana Olsson, Angela Roberts Reeder, Shannon Willis

Maps on pages 22–25 created by Esri. Maps on pages 28–31 created by Mike Boruta.
This book may be purchased for educational, business, or sales promotional use.
For information please write the Special Markets Department at the address or website above.

Library of Congress Control Number 2025044272

Hardcover ISBN: 978-1-58834-811-1
eBook ISBN: 978-1-58834-815-9

Printed in Malaysia
not at government expense
30 29 28 27 26 1 2 3 4 5

TABLE OF CONTENTS

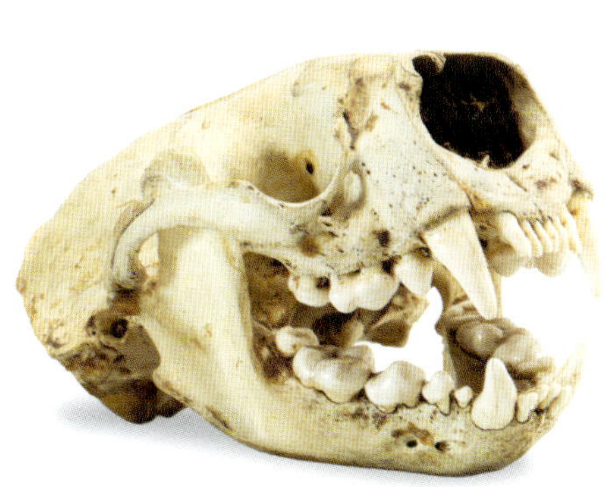

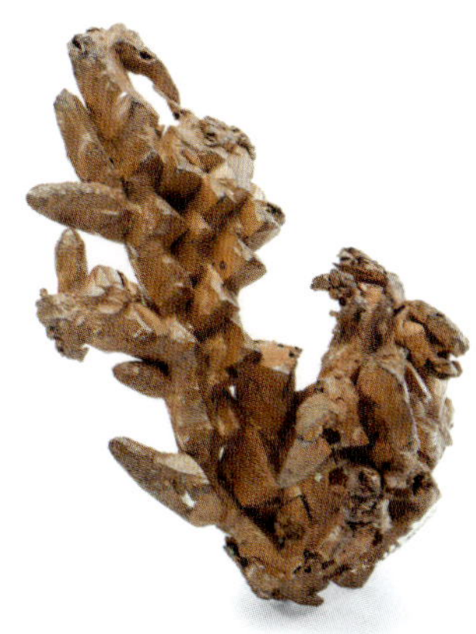

INTRODUCTION

The United States of America and its territories cover more than 3.5 million square miles; stretch from south of the equator to north of the Arctic Circle; and border the Pacific, Atlantic, and Arctic Oceans. These lands and waters are underlain by an ancient and complex geology and host a vast array of ecosystems.

From redwood forests to tropical islands to arctic tundra to shortgrass prairie to deciduous forests to coastal mangroves to oak savannas to icy mountains to rocky deserts, each of these landscapes hosts its own bespoke biology. The extraordinary variety of plants and animals in these places is regularly renewed in the dance of the seasons and bolstered by the appearance of migratory species. The bedrock, rich in minerals, ores, fossils, and gems, has borne witness to the unimaginable forces and seemingly endless time that have birthed and formed the land itself.

The incredibly diverse natural world of the United States of America has supported and nourished people for at least the last 23,000 years and continues to inspire the more than 340 million people who call these lands home today.

In our exhibition ***From These Lands: Sharing Our Natural and Cultural Heritage***, every state and inhabited territory, plus Washington, DC, is represented by at least one intriguing item selected from among the 148 million items in the collection of the National Museum of Natural History. Within the pages of this book, you will find some of our favorites, chosen because we found them beautiful, strange, compelling, informative about a culture, important to science—or all the above. We hope you will share our enthusiasm and will be inspired to make your own discoveries within and about the places you treasure and call home.

I want to thank everyone who supported the development of the exhibition and this book, including the museum's staff, Esri for developing the beautiful and informative maps on pages 22–25, and the Windland Smith Rice Endowment for its generous funding of this and other exhibitions. I also want to thank most especially the people who visit our museum and love this country.

KIRK JOHNSON
SANT DIRECTOR

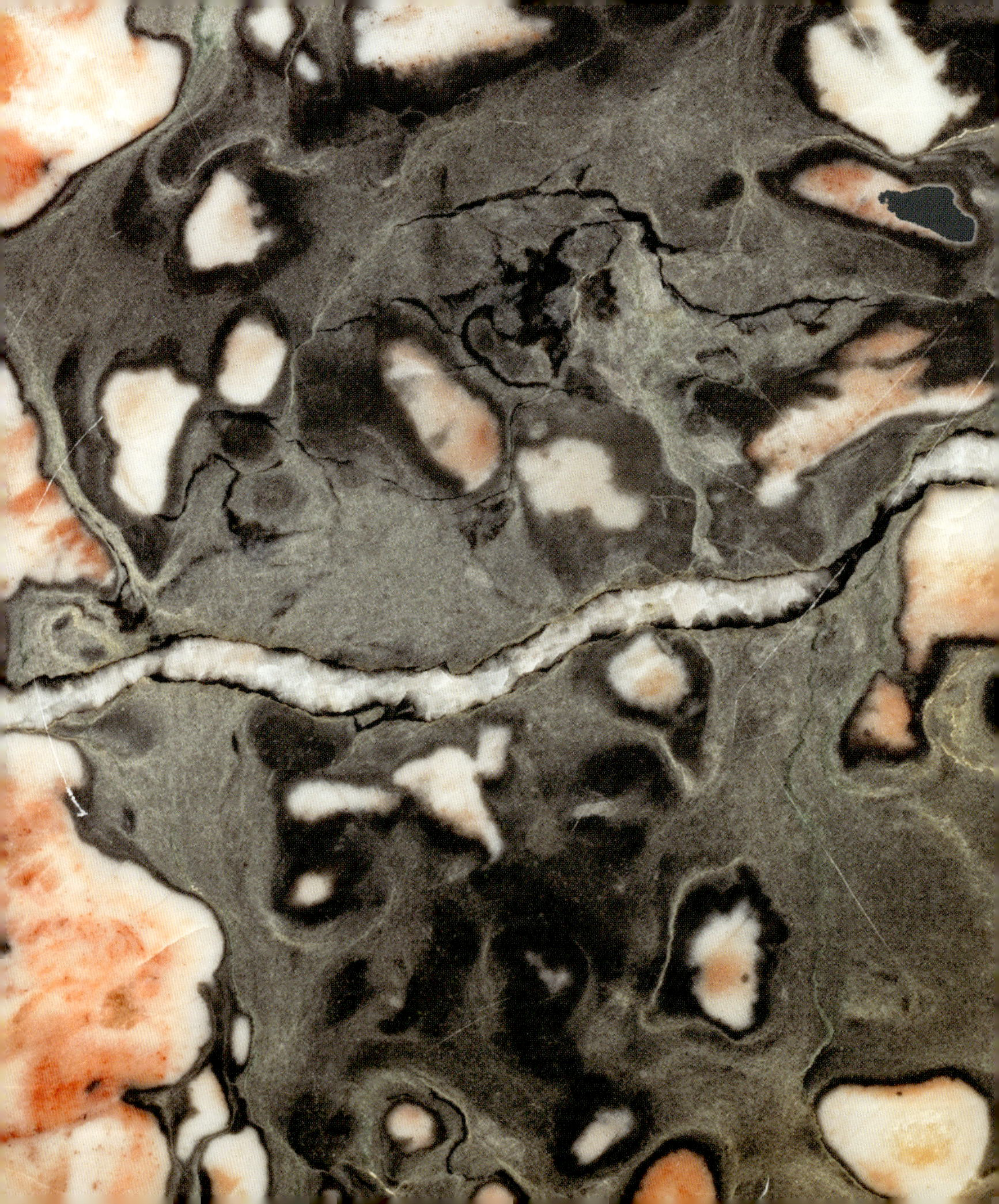

DOCUMENTING THE DIVERSITY OF THE UNITED STATES

From snow-capped mountains to warm ocean waters; from the wide-open plains to the dark recesses of a cave; from massive condors in flight to tiny marine snails—the United States is a study in diverse habitats, ecosystems, and life.

The National Museum of Natural History's collections preserve information about past and present environments and populations. Specimens obtained today create a crucial reference record for future researchers.

Explore just some of the diversity that the museum has documented in its specimens and cultural items over its history.

Insects comprise more than two-thirds of all described species of animals. No group is more diverse.

Insects have evolved an impressive array of colors, shapes, sizes, and lifestyles. There are an estimated 91,000 known insect species in the United States—and many more yet to be discovered!

Besides insects, the 34.5 million specimens in our entomology collection also include arachnids (spiders, scorpions, ticks, and mites) and myriapods (millipedes and centipedes).

Can you find some hidden in these photographs?

Most of the 600,000 bird specimens in our museum are over one hundred years old. They are a priceless record of our country's biodiversity.

Originating from the heyday of exploring expeditions in the late nineteenth and early twentieth centuries, our bird collections preserve information about past and present environments and populations across the country. Researchers can use these collections to understand how biodiversity in the United States has changed over the last hundred years. Some species represented in the museum's collections, such as the Carolina parakeet, were once common in some areas but are now extinct. Specimens in our collections also help with mapping the avian genome. As scientific knowledge expands, the scientific names used may change, such as the cerulean warbler's.

With more than 100,000 species, phylum Mollusca is extremely diverse, with many marine species sporting eye-catching shells.

Of the twenty-three states with marine coastlines, sixteen have official state shells. These shells represent a variety of mollusks, from snails and oysters to scallops and clams. However, mollusks aren't just found in oceans—they also live in terrestrial and freshwater environments in nearly every ecosystem on Earth. You'll find land and tree snails in every state, and the United States is a global hot spot for freshwater mussels, with about three hundred described species (roughly a third of the world's total).

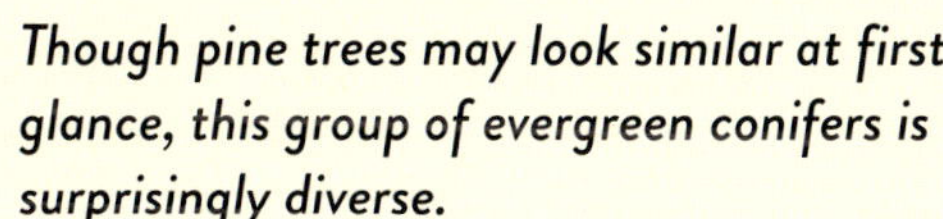

Though pine trees may look similar at first glance, this group of evergreen conifers is surprisingly diverse.

Nearly a third of the world's pine tree species are native to the United States. These cones represent just fourteen of the nearly forty pine species in the country. Every pine species bears unique cones adapted to disperse seeds effectively in their native habitats, whether that's the dry mountain slopes of Nevada, fire-dependent lowlands of South Carolina, or temperate rainforests of California. Pine trees feed and house many other organisms, and humans rely on them for lumber, turpentine, landscaping, and more.

The paleobiology collections contain more than 43 million fossil specimens, including early whales, the earliest flowering plants, ancient insects captured in amber, single-celled organisms, sediment samples—and, of course, dinosaurs.

The mineral sciences collections include more than a million gems, minerals, meteorites, rocks, ores, and drill cores, representing a variety of geological processes and the evolution of Earth and our solar system.

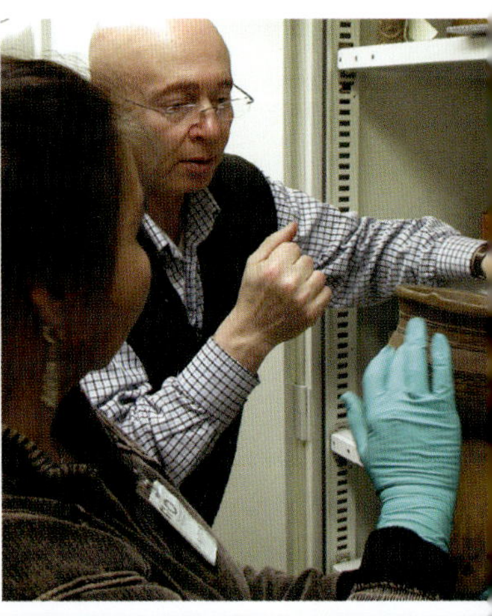

The 3.5 million objects in the anthropology collections include excavated artifacts, culturally important animal and plant remains, cultural artifacts, films, photographs, and sound recordings. These materials represent a rich diversity of cultures and languages from around the world. Following the National Museum of the American Indian Act and Smithsonian policies, we are working to repatriate and return to descendants and their communities ancestral remains and funerary belongings in our care.

Different colors in this map show broad differences in the **elevation** or height of the North American landscape from sea level (green) to mountainous regions (brown and white). Elevation is calculated from sea level (or zero meters).

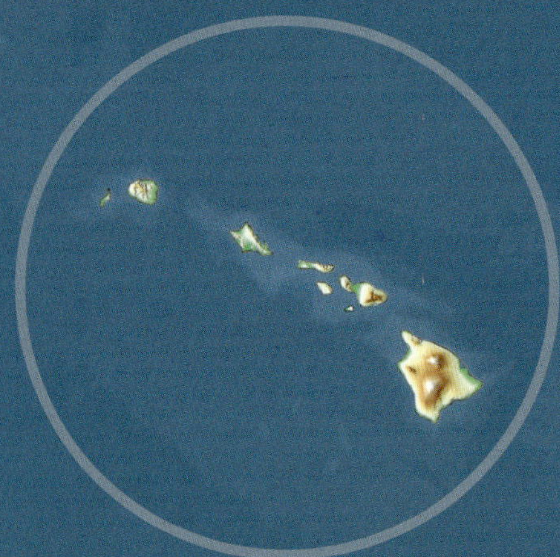

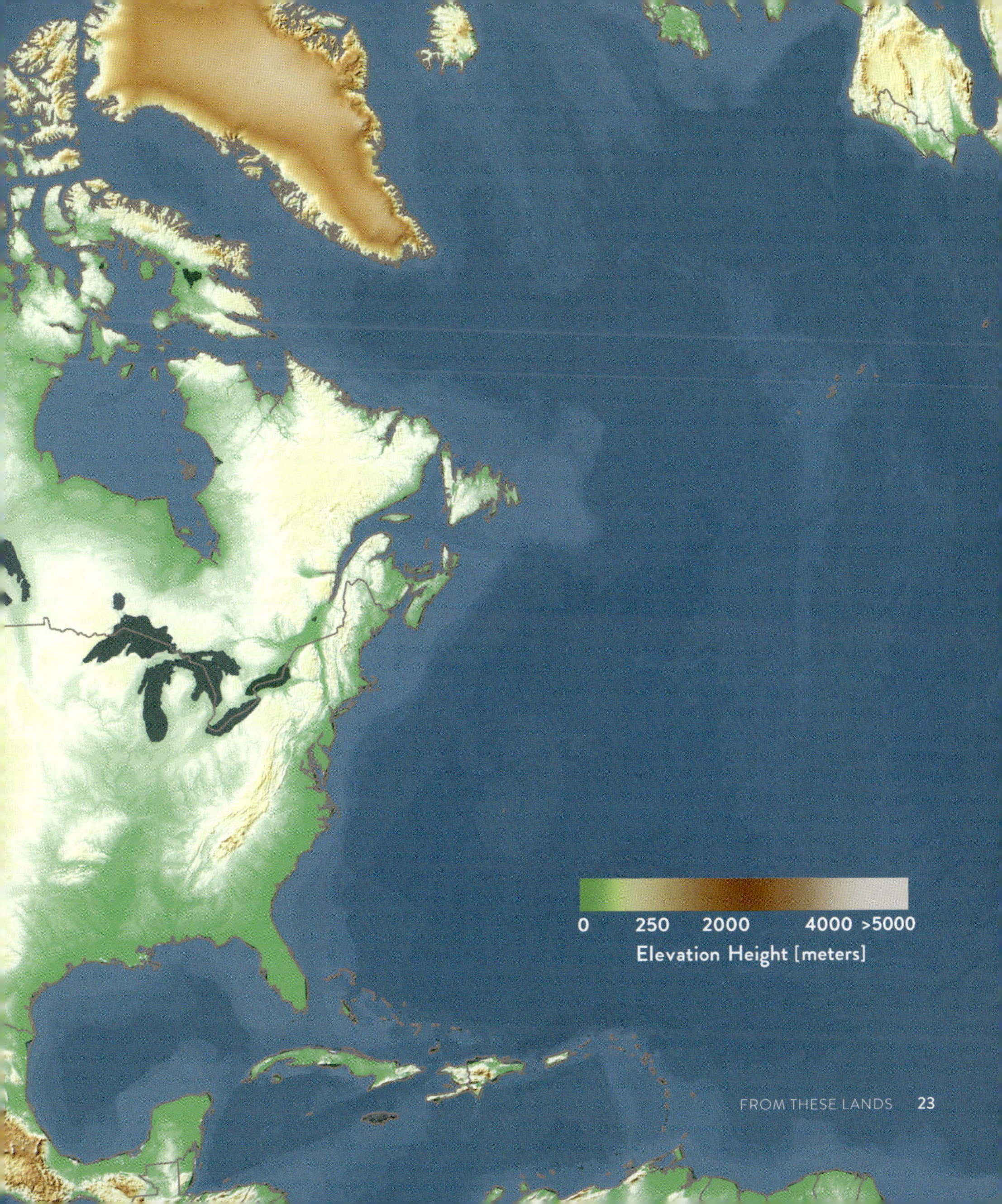
0
250
2000
4000
>5000
Elevation Height [meters]

Biomes are large geographic regions that are home to similar plants, animals, climate, and soil. These regions are often classified by the dominant climate (such as "temperate") and plants ("deciduous forests").

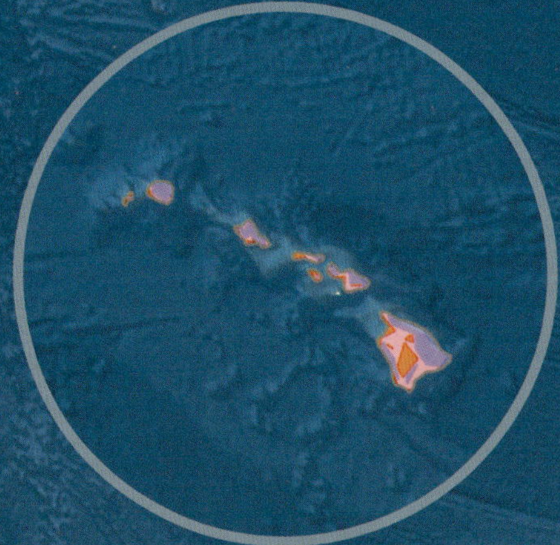

Ice
Desert
Tundra
Grasslands
Rock
Boreal Forest/Taiga
Temperate Deciduous Forests
Mediterranean Woodlands
Temperate Coniferous Forests
Tropical & Subtropical Wet Forests
Tropical Dry & Coniferous Forests

COLLECTIONS FROM THE STATES AND INHABITED TERRITORIES

As part of the Smithsonian's *Our Shared Future: 250* offerings to celebrate and reflect on the 250th anniversary of the US Declaration of Independence, the National Museum of Natural History is hosting the *From These Lands* exhibition to share the natural and cultural heritage of the United States with the public.

Our collections hold the stories of the land, wildlife, history, and people of the United States. Since the Smithsonian's founding in 1846, the museum has collected specimens and cultural items that further scientific research, inspire wonder, invite investigation, spark conversation, and expand perspectives.

In this section, you will find objects and stories from each state and inhabited territory. The objects selected represent the museum's seven science departments: anthropology, botany, entomology, invertebrate zoology, mineral sciences, paleobiology, and vertebrate zoology. Each object provides insight into the history and biodiversity of our great nation. Some will delight and surprise you, whereas others will make you contemplate our nation's past, present, and future.

What will you find from your home?

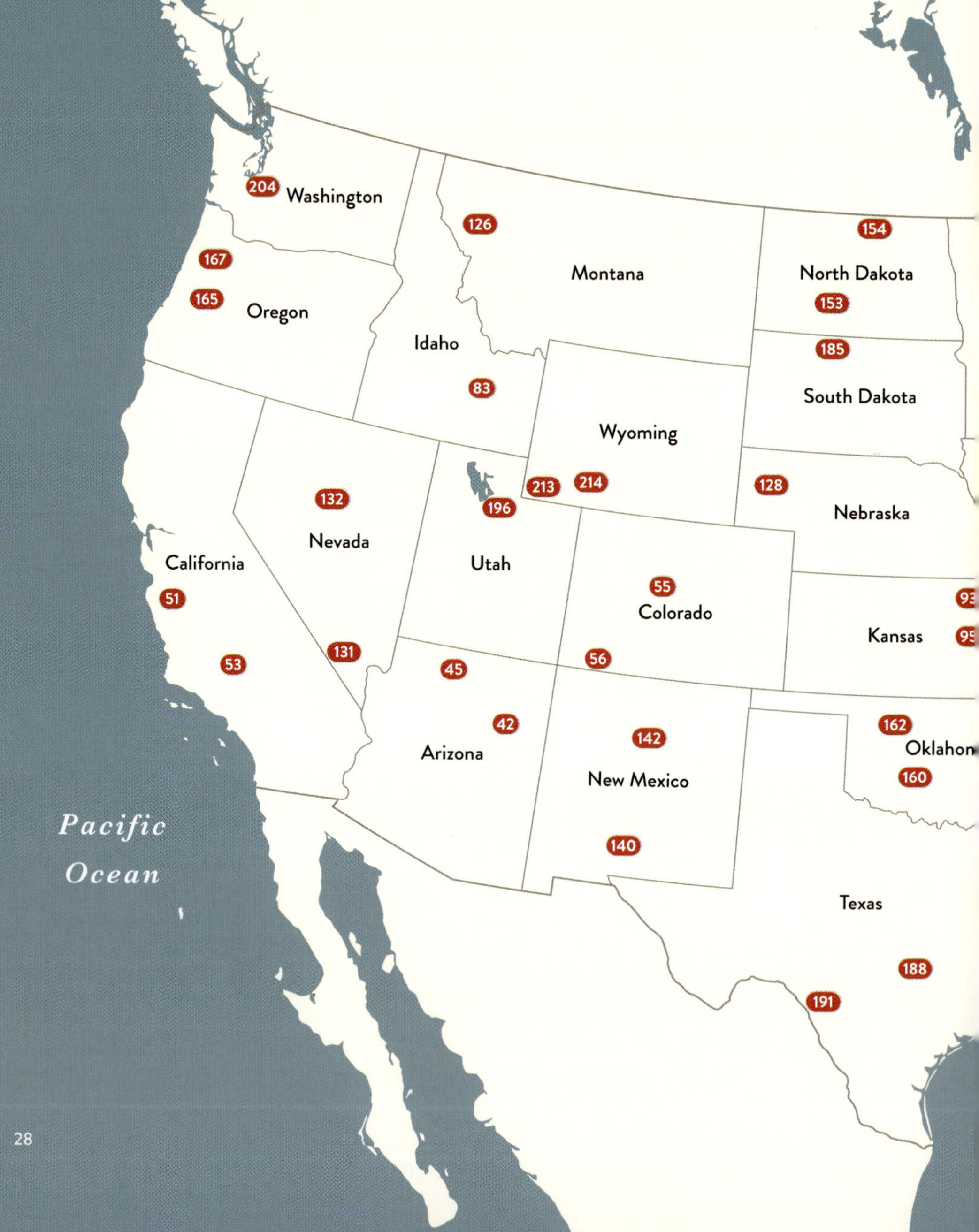
204
Washington
126
154
167
Montana
North Dakota
165
Oregon
153
Idaho
185
83
South Dakota
Wyoming
213
214
128
132
196
Nebraska
Nevada
California
Utah
55
51
Colorado
Kansas
131
53
45
56
42
162
142
Arizona
New Mexico
160
Pacific
Ocean
140
Texas
188
191

nnesota
116
115
208
Wisconsin
Michigan
210
owa
90
Illinois
Indiana
Ohio
87
88
159
124
Missouri
84
96
Kentucky
98
186
Tennessee
Arkansas
46
Mississippi
123
Alabama
34
Louisiana
33
100
103
Maine
105
106
198
Vermont
New Hampshire
134
144
Massachusetts
New York
179
113
61
176
58
Rhode Island
147
Connecticut
Pennsylvania
139
137
169
171
New Jersey
108
66
64
206
69
Delaware
63
203
West Virginia
201
DC
110
Maryland
Virginia
Atlantic Ocean
150
North Carolina
149
South Carolina
74
182
180
Georgia
72
Florida
71
The numbers on the map correspond to page numbers for that object.

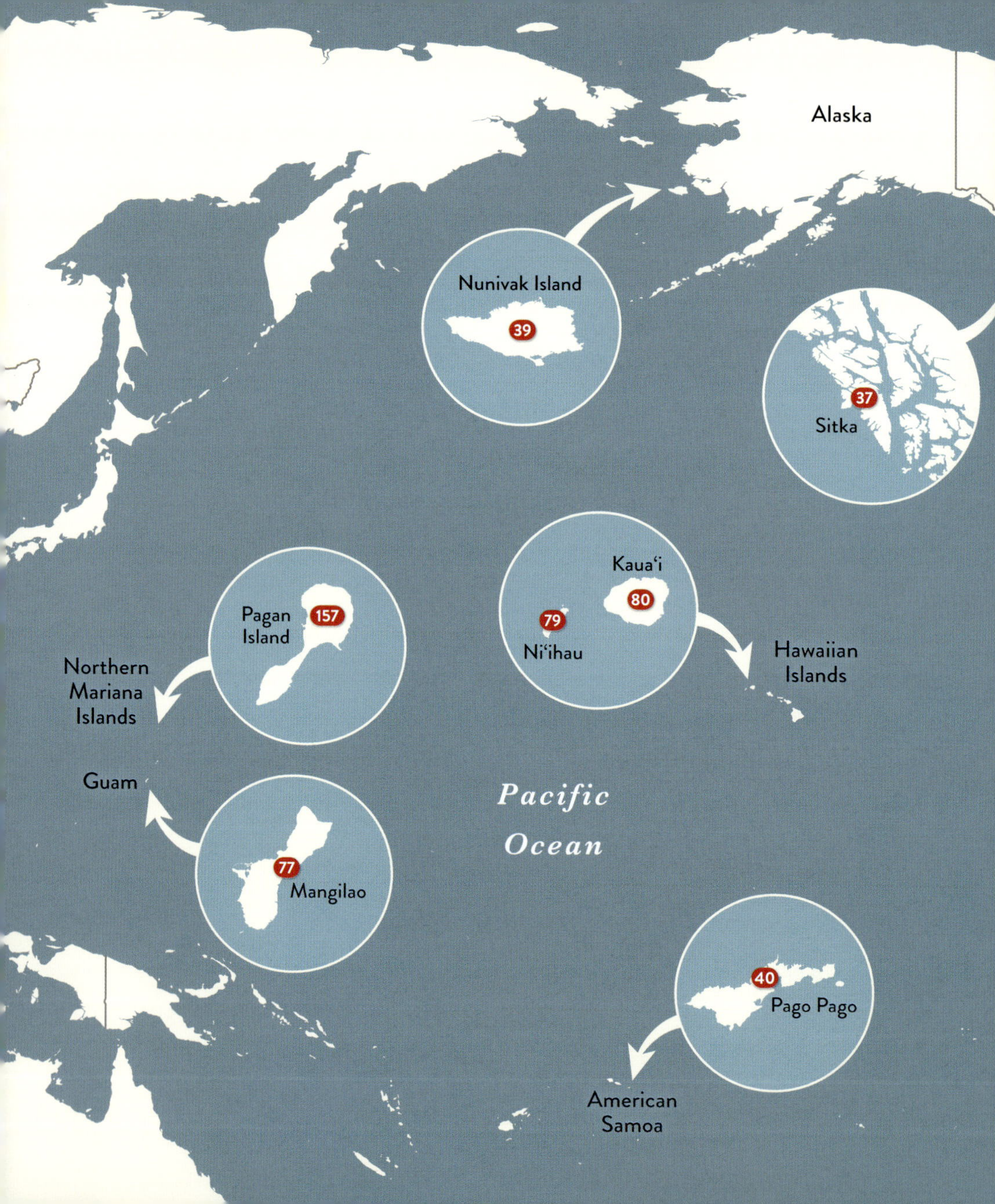
Alaska
Nunivak Island
39
37
Sitka
Kaua'i
80
79
Ni'ihau
Hawaiian Islands
Pagan Island
157
Northern Mariana Islands
Guam
77
Mangilao
Pacific Ocean
40
Pago Pago
American Samoa

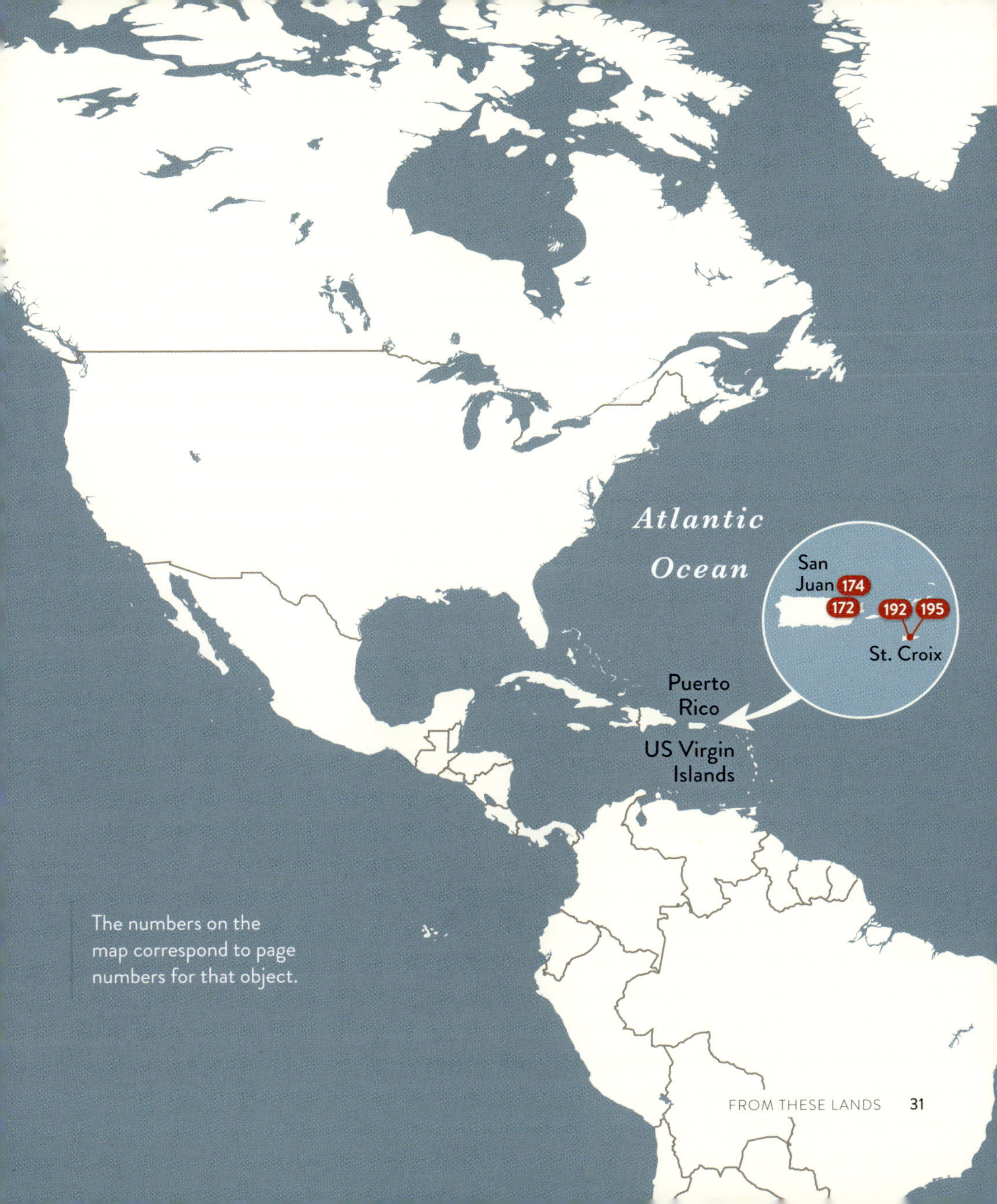
Atlantic
Ocean
San
Juan
174
172
192
195
St. Croix
Puerto
Rico
US Virgin
Islands
The numbers on the
map correspond to page
numbers for that object.

ALABAMA

Johnstone's Junonia

Scaphella junonia johnstoneae

Collected in 1951 from Petit Bois Pass, Alabama

USNM MOL 598100

Junonias are a rare beach find—and the subspecies Johnstone's junonia is even more so. The official state shell of Alabama, this carnivorous marine snail lives only in waters ten to forty-five miles (16–72 km) off Alabama's coast between Bayou La Batre and Gulf Shores. Junonias live on the ocean floor in waters 95 to 415 feet (29–126 m) deep where they can grow up to 4.5 inches (11 cm) long. It takes a strong storm or hurricane to bring these shells to the beach naturally, but many are taken as accidental bycatch during commercial shrimp trawling. Johnstone's junonia looks similar to other junonias except for its light brownish orange coloring.

ALABAMA

Eastern Diamondback Rattlesnake

Crotalus adamanteus

Collected in 1976 from Washington County, Alabama
USNM 218659

Eastern diamondback rattlesnakes are North America's largest venomous snakes. Adults can grow up to eight feet (2.4 m) long! They are found in grasslands and forested habitats throughout the southeastern United States, from southeastern North Carolina down into Florida and along the gulf coast to Louisiana. They are ambush predators and wait coiled and camouflaged for their prey to cross their path before striking. Rattlesnakes vibrate the rattles on their tails when they feel threatened. When you hear that sound, it's best to give them plenty of space; rattlesnake venom can be fatal to humans.

ALASKA

Wéix' S'áaxw (Sculpin Hat)

Original collected in 1884 from Sitka, Alaska
New hat made in 2019 for the Sitka Kiks.ádi clan
NMNH E74339, NMNH E74441

In Tlingit culture, hats such as this are sacred objects, or *at.óow*, imbued with the spirits of ancestors and used in dancing and ceremonies. The sculpin, or *wéix'*, depicted in this hat is a fish abundant along the Pacific Northwest coast and a crest of the Kiks.ádi clan. In 2012, the original hat was deemed too fragile for ceremonial use, so Smithsonian staff and members of the Kiks.ádi clan collaborated to digitally scan the original hat and create an exact replica, following Tlingit cultural protocols and using materials sourced from Alaska. The Tlingit brought the new hat home in 2019, where it was consecrated as a sacred object they could dance with once again. A second replica remains in the museum for educational use.

ALASKA

***Kaugpaq Tugkarer* (Walrus Tusk)**

Carved c. 1929
Collected in 1957 from Nunivak Island, Alaska
Gift of Mrs. Edward Darlington Jones
NMNH E394454

This carved walrus tusk illustrates the strong ties between walruses and the Cup'ig who live on the coasts of the Bering Sea. For thousands of years, the Cup'ig have relied on walruses for food and material resources, from skin boats and drums to tools and crafts. Indigenous artists would showcase their skills and pay respect to the walrus through the creativity and complexity of their work on these tusks. This carving portraying Alaskan wildlife was made to be sold to visitors (including museum collectors). The Indigenous knowledge etched in the ivory speaks to the Cup'ig deep understanding of walruses—notice the walrus preying on a seal, an uncommon behavior but one that hunters knew to look out for.

AMERICAN SAMOA

Historic and Contemporary *Siapo* (Bark Cloth)

Historic *siapo* collected in 1890 from Pago Pago, American Samoa
Contemporary *siapo* made in 2014 by Su'a Uilisone Fitiao (b. 1963?)
Gift of the artist
NMNH E151409 *(right)*, E434512 *(left)*

Known as *siapo* in American Samoa and *kapa* in Hawai'i, Pacific Island bark cloth has always been made to be used. Traditionally, it could be found in ceremonial and everyday dress, uniforms, shrouds, and room dividers. Both the historic rectangular cloth and the contemporary circular cloth were painted freehand in the *siapo mamanu* tradition, featuring traditional motifs integral to Samoan culture. The techniques and patterns of siapo would have been lost without the commitment of artists keeping the practice alive. Today, *siapo* is worn in formal and casual clothing; decorates the home; and adorns purses, hats, and jewelry.

ARIZONA

Petrified Tree

Agathoxylon arizonicum

Lived 219-213 million years ago
Donated in 1991, collected from Navajo County, Arizona
USNM PAL 455192

Each part of this fossil tells a story about its formation. About 216 million years ago, a decaying log got stuck in a river channel, perhaps during a flash flood. The uneven margins along the tree's outer edge show that it had decayed before being preserved. Intense water flow swept together a mix of other stones, which encircled the tree. These came from older rocks that already contained marine fossils. Once the log was buried, groundwater seeped into it and coated and filled the tree's original cell structures with silica. Elements such as iron and manganese trapped in the silica crystals produced a rainbow of colors, from purple to yellow to smoky gray quartz. You can find many similar examples of petrified wood protected at Petrified Forest National Park in Arizona.

13

ARIZONA

California Condor (Male #413)

Gymnogyps californianus

Collected in 2013 from Kaibab Plateau, Arizona
USNM 658157

The massive California condor evolved to scavenge the carcasses of now extinct mammals, such as giant ground sloths and mastodons. More recently, lead poisoning (from the ammunition used to hunt game), predation, and injuries caused by powerlines threatened this species until only twenty-two individuals were alive in 1982. Because of the US Fish and Wildlife Service's (FWS) California Condor Recovery Program, more than 550 of these birds are now thriving in the Southwest. Each is tagged at hatching, tracked throughout its life, and collected and stored after death. In 2017, FWS donated a few specimens to the museum for research and display. Researchers can use the data collected during and after #413's life to better understand the California condor's life cycle, behavior, and ecology.

ARKANSAS

Bauxite

Sedimentary rock

Collected prior to 1958 from Saline County, Arkansas
NMNH 116537-74

It's no wonder that bauxite is the state rock of Arkansas, as this state was once the largest source of bauxite ore in the United States. Many states pick their symbolic rocks for their economic value—and bauxite is the primary source of aluminum, a metal with many commercial and industrial uses. Aluminum is one of the least expensive metals to buy and use today, but in the 1800s, it was worth more than gold. In 1886, a reliable, quick, and inexpensive process was invented for extracting aluminum from ores (such as bauxite). Now, aluminum is used in everything from space vehicles to soda cans!

ARKANSAS

Conehead Katydid

Neoconocephalus sp.

Collected in 1968 from Devil's Den State Park, Arkansas
USNM ENT 01978917

Coneheads are a genus of katydids with (you guessed it) a cone-shaped head. There are 130 species of coneheads, and they live across the United States and northern Mexico. Coneheads are usually green, but some may be tan or yellow. They blend into the high grasses in which they live, eat, and reproduce. Their strong jaws allow them to easily eat grass seeds—and bite anything that tries to handle them! Females, such as this specimen, have a long, sharp ovipositor extending from the end of their abdomen, which is used to cut into plant tissue and deposit fertilized eggs into the plant. For their part, males "sing" during mating season by rubbing their wings together in the evenings. Each species has its own song that allows them to find the right partner.

CALIFORNIA

Benitoite Mineral and Brooch

$BaTiSi_3O_9$

Specimen *(right)* acquired in 1974; collected from San Benito County, California
Gift of J. J. Trelawney, USNM 137644

Brooch *(left)* acquired in 2023
Purchase made possible by Smithsonian Gem and Mineral Collectors, USNM G11821

Benitoite, which ranges in color from violet-blue to almost colorless, became the state gem of California in 1985. It forms under low temperatures in high-pressure subduction zones where tectonic plates collide and one sinks into Earth's mantle underneath another.

Rarer than gold, softer than a sapphire, and more brilliant than diamonds, gem-quality benitoite has only been found in a small area in the Diablo Range between Los Angeles and San Francisco, where it was discovered in 1907 by prospector James M. Couch. Small amounts of lesser quality benitoite have been found in other parts of the world, but the Diablo Range area has been mined out.

Skull

Hind limbs

CALIFORNIA

Early Pinniped

Enaliarctos mealsi

Lived 25–24 million years ago
Collected in 1975 from Kern County, California
USNM PAL 374272

An early relative of modern seals, sea lions, and walruses, this fossilized pinniped helps pinpoint when this group of mammals began transitioning from land to water about 34–28 million years ago. Museum preparators carefully exposed both sides of this ancient marine mammal while keeping it in its original position, preserving information about what happened between death and burial. Scientists and illustrators could then reconstruct the animal to study its anatomy for clues about its lifestyle and evolutionary history. The shape of the hind limbs indicates that this pinniped moved easily both on land and in water, like a sea lion, suggesting it might have spent most of its time near shore.

COLORADO

Rhodochrosite with Tetrahedrite and Quartz

$MnCO_3$, $(Cu,Fe)_{12}Sb_4S_{13}$, and SiO_2

Collected in 1992 from Park County, Colorado
Gift of Washington A. Roebling
USNM R19750

The deep rose-red of its purest forms makes rhodochrosite a brilliant choice for Colorado's state mineral. The state derives its name from the Spanish word *colorado*, meaning "colored red." Rhodochrosite gets its red and pink colors from manganese trapped in the crystal lattice of this mineral. How does this happen? Water deep in the Earth is heated by magma and then flows through fissures or cracks in rocks, depositing minerals—such as rhodochrosite, tetrahedrite, and quartz—that crystalize in these cracks as the water cools.

Rhodochrosite was not considered commercially valuable and was discarded as waste when found during early silver mining. Specimens from the Sweet Home Mine in Alma, Colorado, such as this one, are among the best examples of rhodochrosite in the world.

COLORADO

Fossilized Paper

Acquired in 1913, collected from La Plata County, Colorado
USNM PAL 799125

Not all fossils are ancient. We don't have an exact date for this specimen, but sometime between the mid-1800s and 1913, a printed piece of paper was blown into a calcium-rich spring, which preserved it down to the letter. Though the text is backward because it's an impression, it's possible to make out some of the words, such as *governor*, *business*, *unhooked*, *opportunity*, and *Senegambian*. The columns of text separated by a vertical line call to mind newspaper or magazine layouts, making this some very old news indeed.

CONNECTICUT

Almandine Garnet

$Fe^{2+}_3Al_2Si_3O_{12}$

Acquired in 1927, collected from Fairfield County, Connecticut
Gift of Frederick A. Canfield
USNM C2731

Formed in high-temperature and high-pressure environments—and typically found embedded in metamorphic rocks—almandine, a member of the garnet group, is literally and figuratively multifaceted. Its complex crystals, which have forty-eight faces, are both attractive and functional. Gem-quality garnets such as almandine seem to blaze like small, hot coals. Even rough garnets have value. They're harder than steel, making them useful as abrasives in grinding wheels, sandpaper, and saws. In fact, these industrial uses are why almandine is Connecticut's state mineral.

CONNECTICUT

Giant Laphria

Laphria grossa

Collected in 1937 from Litchfield County, Connecticut
USNM ENT 01978195

That large bumblebee buzzing around might not have a stinger—in fact, it could be a species of robber fly, whose black and yellow bands and body shape closely resemble a queen bumblebee. Robber flies, like the giant laphria, lie in wait on tree branches or other perches to ambush insects, primarily bees and beetles, in flight. A robber fly will pierce the body of its prey and kill it with its paralyzing saliva before landing to suck out the fluids and soft organs, leaving only the hard outer shell. There are over seven thousand species of robber flies worldwide. The giant laphria can be found in deciduous forests throughout the eastern United States and Canada.

DELAWARE

Channeled Whelk

Busycotypus canaliculatus

Collected in 1975 from Cape Henlopen State Park, Delaware

USNM MOL 806855

Found from southern Massachusetts to northern Florida, the channeled whelk is a large predatory sea snail that lives in sandy oyster beds, mud, and sand flats near the ocean shore. They are nocturnal and eat clams, mussels, and oysters. Their main predator is the blue crab. These whelks are named for the conspicuous channel that follows the spiraling growth of their smooth, cream-colored shell, which ranges from five to eight inches (12.7–20.3 cm) in length. The greenish-brown fuzzy material around the opening, or aperture, of this specimen is the remnants of the organic outermost coating, called the periostracum, of the shell found on many mollusks.

DELAWARE

Horseshoe Crab

Limulus polyphemus

Collected in 2021 from Delaware Bay, Delaware
Gift of the Delaware Museum of Nature & Science
USNM IZ 1548559

Horseshoe crabs aren't crabs. They're not even crustaceans! Their closest relatives are spiders and scorpions. From May through early June, hundreds of thousands of horseshoe crabs converge on Delaware Bay beaches to mate and deposit their eggs, attracting both nature enthusiasts and migrating shorebirds. Horseshoe crabs have been around for more than 445 million years and survived multiple mass extinctions, but they are now threatened by human activities. Fishermen still use them as bait, and farmers used them as fertilizer until synthetic fertilizers became available in the early twentieth century. Since 1982, the biomedical industry has harvested their blood for a special enzyme that detects bacterial toxins in patients, drugs, and medical devices. A synthetic alternative approved in 2024 could save both horseshoe crabs and people.

DISTRICT OF COLUMBIA

Wild Rice

Zizania aquatica

Collected in 1958 from Theodore Roosevelt Island, District of Columbia
US-00371688

Wild rice once thrived along the District of Columbia's rivers and was a dietary staple for the Piscataway and Nacotchtank peoples living in the area. However, as development transformed the region's watershed, its freshwater wetland habitat disappeared. Today, wild rice is making a comeback along the Anacostia River as conservation organizations and community members reenvision what an urban river can be. Wild rice provides a habitat among its roots and shoots for fish, frogs, migratory birds, insects, and invertebrates. Its seeds are also a vital food source for migratory birds and other wildlife. In addition, these grasses filter freshwater tidal marshes, trap sediments and pollutants, help control storm surges, and store carbon.

Zizania aquatica L. (typical)
Roosevelt Island = Theodore Roosevelt Island
Lindsey Kay Thomas, Jr. July 20, 1966
Zizania aquatica L. var. aquatica
P. M. PETERSON 1996
Smithsonian Institution
DISTRICT OF
Zizania aq tica L. det. FAM
In shallow water (mud at low tide) on
southeast shore of Roosevelt Island.
Annual; the plant to about 3 m. tall.
Staminate florets persistent, the
fruits fugacious. Coll.McClure 10X58
Sheet 2/1
00371688
2313300

DISTRICT OF COLUMBIA

The District of Columbia is an urban environment full of natural areas and green spaces that support a wide variety of insects and other arthropod life, like those shown in this image. Some are easily spotted, such as vibrant painted lady butterflies. Others, like the appropriately named flea beetle, might require a magnifying glass to see. You can also find spiders, such as the spined micrathena in its orb-shaped web, but remember, spiders aren't insects—they're arachnids.

Come summertime, it's hard to miss the chorusing katydids, crickets, and cicadas—just listen for their trills, rasps, buzzes, and chirps. As you listen and watch out for cicadas, keep an eye out for eastern cicada killers, a type of wasp that preys on cicadas twice their size.

FLORIDA

Horse Conch

Triplofusus giganteus

Collected prior to 1892 from Monroe County, Florida
USNM MOL 124855

Florida's state seashell is the protective hard shell of an extremely large predatory sea snail—the horse conch. These gastropods live in subtropical and tropical shallow waters off the Atlantic coast from North Carolina to Mexico's Yucatán Peninsula. At adulthood, they weigh eleven pounds (5 kg) on average and their shells can reach twenty-four inches (61 cm) in length, making them the largest gastropods in US waters! Their shells are bright orange when they are young but become more muted to grayish white as they mature, and their soft parts are also bright orange. They use their powerful large foot to trap and smother their prey—primarily other large sea snails. Horse conch populations are in decline due to overharvesting, ocean acidification, and infrequent spawn rates.

FLORIDA

Carolina Parakeet (extinct)

Conuropsis carolinensis

Collected in 1896 from Osceola County, Florida
USNM 220629

The Carolina parakeet was the only parrot native to North America north of the Rio Grande River. The last confirmed sightings of a Carolina parakeet in the wild occurred in the 1930s, and the species probably was extinct by the early 1940s. Much of what we know about this species comes from written accounts prior to the twentieth century and contemporary studies of museum specimens, which help scientists understand the Carolina parakeet's evolutionary history and biology. The Carolina parakeet once lived in Florida, the coastal areas of Mississippi and Alabama, along the Ohio and Mississippi rivers south of central Iowa, and into the southeastern Great Plains. Although deforestation and hunting played major roles in their decline, the cause for their final extinction is still unknown.

GEORGIA

Cave Salamander

Eurycea lucifuga

Collected in 1987 from Walker County, Georgia
USNM 467977

The southern Appalachian Mountains have the highest diversity of salamander species of any place on Earth, thanks to the region's varied elevation, cool streams, and moist forests. Cave salamanders are particularly partial to this region of the United States, where there are plenty of caverns, rock faces, and springs. Even in the darkness, cave salamanders stand out. Adults are typically bright orange with dark spots, and they can grow to be just under seven inches (18 cm) in length. Like many of their closest relatives, cave salamanders secrete a noxious substance when threatened, deterring would-be predators. They also hunt mainly at night, staying hidden in the dark recesses of their caves and rocky crevices during daylight hours.

467977

GUAM
JFL

GUAM

Traditional CHamoru Tools

Joaquin Flores Lujan (1920–2015)
Made in 1996 in Mangilao, Guam
Transfer from the National Endowment for the Arts Folk Arts Program
NMNH E428722 *(left)*, E428718 *(center left)*,
E428727 *(center right)*, E428716 *(right)*

There was no significant source of metal in Guam until contact with Europeans, who subsequently initiated trade with the island and its residents in 1521. For centuries, Guam's traditional blacksmiths have blended colonial influences with CHamoru ingenuity as they developed techniques for using salvaged and imported metal. However, economic changes and societal shifts after World War II nearly brought the practice to an end. By the 1970s, Joaquin Flores Lujan, a US Navy welder and immigration officer, was the last traditional *herrero* (blacksmith). He'd promised his father to keep the practice alive for future generations and went on to train more than a dozen apprentices. The traditional tools pictured here (fishing spear, hoe, betel nut scissors with wooden stand, and machete) are made from carbon steel, wood, and copper.

HAWAIʻI

Pupu Shell Lei

Acquired in 2003 from Niʻihau, Hawaiʻi
Bequest of Betty Jane Henderson
NMNH E430782

Tiny pupu (or snail) shells are the official lei material of the Hawaiian island of Niʻihau. This three-stranded lei contains at least five different species of sea snails, including momi (*Euplica varians*), kahelelani (*Homalopoma verruca*), and the Hawaiian top shell (*Turbo sandwicensis*). Some, like the Hawaiian top shell, can only be found in Hawaiʻi, whereas others range more broadly throughout the tropical Pacific and Indian Oceans. Creating leis from seasonal and local materials and gifting them to friends and family during special occasions is a time-honored tradition and beautiful expression of Kānaka Maoli, or Native Hawaiian, culture and community.

HAWAIʻI

Cliff-hanging Carnation Relative

Schiedea waiahuluensis

Cultivated from seed collected in 2022 from Kauaʻi, Hawaiʻi
US-04162545

There are thirty-six species in the genus *Schiedea* within the carnation family—and all of them are found only in Hawaiʻi. This species is among the most rare and hard to find, as it only grows high up on the steep and inaccessible cliffs of the Waiahulu Valley on the island of Kauaʻi. Threatened by feral goats, which were introduced in the late 1700s, and by invasive plants, this species is critically endangered. Researchers at the National Tropical Botanical Garden first spotted this unique species in 2021 when monitoring Waimea Canyon with drones. They used a drone equipped with a specialized claw to collect cuttings and seeds. Cultivated from those seeds for future scientific study and conservation efforts, this specimen was among those used to officially describe the species as new to science in 2024.

UNITED STATES
3770711
NATIONAL HERBARIUM
Image No.
04162545
Plants of the Hawaiian Islands
ISOTYPE
Caryophyllaceae
Schiedea waiahuluensis W.L. Wagner, Weller, B. Nyberg & A.K. Sakai
Cultivated in greenhouse at University of California, Irvine.
Plant originally discovered and a fruiting branch was collected by NTBG drone program team. A flowering specimen was collected via drone collecting arm [Waimea District, north-facing dry cliffs above Waiahulu, 767 m elev., 29 Mar 2022, B. Nyberg et al. BN023 (PTBG)], a plant was subsequently cultivated from seed harvested from this collection.
S. G. Weller and A. K. Sakai 1172
12 April 2024
United States National Herbarium (US)
Smithsonian Institution

IDAHO

Lava Bomb

Igneous rock

Collected prior to 1903 from the Snake River Plain, Idaho
NMNH 75261

The United States has more volcanoes (both active and inactive) than any other country. When they erupt, some ooze lava and others eject fast-moving clouds of gas, ash, and rocks. Volcanoes between these two extremes experience explosive eruptions of hot gases, which can launch molten or partially molten lava "bombs" and other rock fragments into the air. Lava bombs are larger than 2.5 inches (6.4 cm) in diameter. Some are stretched like taffy mid-flight, whereas others flatten when they slam into the ground. These rocks pile up, creating a steep cone of ejected pyroclastic material around the volcano's vent. The eruptions that produced this rock occurred sometime between 15,000 and 2,000 years ago.

ILLINOIS

Cricket Frogs

Acris blanchardi

Collected prior to 1873 from Union County, Illinois
USNM 8178

Cricket frogs were once the most common amphibian in Illinois, but their populations have declined. Scientists in the early 2000s suspected that certain hormone-altering chemicals, such as DDT, were to blame, because the frogs live near slow-moving streams, ponds, and ditches that are susceptible to runoff. They examined specimens collected from different counties and decades, using the museum's pre-1873 frogs as a baseline. In areas and time periods when these chemicals were prevalent, cricket frogs had skewed sex ratios and there were more individuals with both male and female sex organs. These factors reduced their breeding success, which led to fewer frogs.

244295

ILLINOIS

Polyphemus Moth

Antheraea polyphemus

Collected from Illinois, year and exact location unknown
USNM ENT 01978114

The giant cyclops Polyphemus from Homer's epic *The Odyssey* lends his name, but not his man-eating tendencies, to this moth due to its size and the large eyespots on its hindwings. The Polyphemus moth has an average wingspan of six inches (15.2 cm) and is found throughout North America. Its eggs can be seen on the leaves of American elm, birch, and willow trees both in the early spring and late summer. As a caterpillar, it eats thousands of times its weight from a wide variety of plants, but for the one week in its adult form, it doesn't eat at all! This male specimen used its large bushy antennae to detect the reproductive pheromones emitted by unmated females.

INDIANA

Crinoid

Barycrinus hoveyi

Lived 358–340 million years ago
Acquired c. 1960s from Indiana, exact location unknown
USNM PAL 351874

Fossils of marine plants and animals end up far from today's coastal shores because the world was very different millions of years ago. The continents took millions of years to move into their current positions. In addition, time periods with high atmospheric carbon dioxide levels and a warmer climate, such as the Carboniferous Period (359–299 million years ago), also had higher sea levels.

During this period, much of the Midwest was under a tropical shallow sea. Though they resemble plants, crinoids are actually animals related to sea stars. They have been around for about 450 million years, but they're much less abundant and diverse today than they were during their heyday.

IOWA

Mussels, Drill Bits, and Mother-of-pearl Buttons

Megalonaias nervosa (mussels), *Reginaia ebenus* (buttons)

Mussels collected in 1907 from Dubuque and Muscatine Counties, Iowa
Buttons created in early 20th century
USNM MOL 746124 *(top left)*, MOL 676959 *(top right)*, MOL 854150

Before the widespread use of plastics, buttons were often made of natural materials, including nacre or "mother of pearl" from the abundant freshwater mussel shells of the Mississippi River. When German immigrant John Frederick Boepple (1854–1912) opened a button factory in Muscatine, Iowa, in 1891, it kicked off an intense period of mussel harvesting that lasted until the 1950s, when the industry declined due to the collapse of the mussel population, the pollution of the river, and the rise of plastics. The United States is home to about three hundred freshwater mussel species—roughly a third of the global total—but most are threatened, endangered, or in decline.

676959

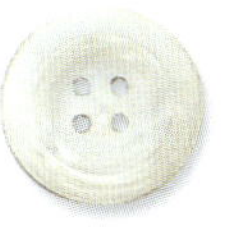

Uniquad

KANSAS

Admire Meteorite and Olivine var. Peridot

Pallasite, PMG and $(Mg,Fe)_2SiO_4$

Collected from Kansas, year and exact location unknown

Olivine gift of Charles Ellias

USNM 1635-D *(left)*, G11416-00 *(right)*

Pallasites are a rare type of stony-iron meteorite that form when two asteroids collide, trapping olivine crystals in a nickel-iron matrix. Olivine is a major component of all rocky planetary bodies—planets, moons, meteorites, comets, and stardust. Its crystals hold clues to the formation of our own planet—and potentially the origins of life. Meteorites such as this one and the minerals they contain provide clues to help us understand how solid bodies collide to build planets and about the birth of our solar system following the explosion of a dying star. The olivine contained within can sometimes be cut and polished into beautiful gems.

KANSAS

Galena

PbS

Acquired in 1974, collected from Kansas, exact location unknown

USNM 134994

Galena is the state mineral of Kansas, Missouri, and Wisconsin! It is such an important source of lead that several towns across the country bear its name. Galena occurs around the world, but in the United States important deposits of galena are found along the Mississippi River. Galena is formed when hot, mineral-rich waters move through fractures in Earth's crust. As the water cools, the minerals crystalize into solid deposits. Galena is often found with other minerals, such as the glittering pyrite seen here and fluorite (page 134).

KENTUCKY

Red-legged Buprestis

Buprestis rufipes

Collected in 1965 from Bourbon County, Kentucky
USNM ENT 01978514

The red-legged buprestis is a wood-boring beetle found in the southern and eastern United States. Although easily identified by the bright orange-yellow stripes and patches on its metallic green elytra (the hardened coverings over its wings), this beetle is rarely encountered, and little is known about it. The females deposit their eggs directly into the wood of dead hickory, beech, chestnut, oak, and tulip trees. Once the larva hatch, they tunnel and feed on the wood for years before emerging as adults.

KENTUCKY

Gypsum Stalactite

$CaSO_4 \cdot 2H_2O$

Acquired in 1892, collected from Mammoth Cave National Park, Kentucky
NMNH 68155

Mammoth Cave is the longest known cave system on Earth, with over 420 miles (676 km) of mapped caverns. It is a testament to the power of water to shape spectacular chambers and structures. The slight acidity of rainwater and groundwater dissolves tunnels in Kentucky's limestone.

At the same time, mineral-filled water forms stalagmites, stalactites, straws, flowstones, and evaporites as it drips and pools inside the caverns. Several pools and streams flowing through the caves support fish, crayfish, shrimp, snails, and thirsty bats.

LOUISIANA

Kudzu

Pueraria montana var. *lobata*

Collected in 1963 from Iberia Parish, Louisiana
US-02343884

East Asian kudzu was first displayed in the United States at the 1876 Centennial Exposition in Philadelphia to highlight its ornamental potential. From the 1930s to 1950s, the Soil Conservation Service (now the Natural Resources Conservation Service) promoted kudzu to control erosion. It worked too well, though, becoming known as "the vine that ate the South." When humans introduce species to new places, it can provide new opportunities for those organisms to thrive. If in these new places they have few diseases or predators, introduced species can quickly take over, upending native ecosystems. Kudzu is considered invasive in the South, where it takes over entire landscapes, outcompeting and literally smothering native plants. It can grow a foot (0.3 m) a day, and mature vines can be one hundred feet (30 m) long.

Herbarium of the University of Southwestern Louisiana
PLANTS OF LOUISIANA
16085. Pueraria lobata (Willd.) Ohwi
flowers fragrant, purple (standard yellow centrally). leaves glaucous below. high climbing vine.
2422075
Edge of woods, Avery Island.
Parish: IBERIA
Date: 21 June 1963
John W. Thieret
Image No.
02343884

LOUISIANA

Shrimp

Glyphocrangon longirostris

Collected in 2000 and 2011 from off the coast of Louisiana
USNM IZ 1022717 *(top)*, IZ 1547396 *(bottom)*

Museum scientists study specimens of marine life collected before and after the *Deepwater Horizon* oil spill in 2010 to understand its impact on deep-sea life. This information can inform conservation and restoration efforts. The effects of the largest oil spill in US history on surface water and coastal habitats were well documented, but about 20 percent of the oil sank to the seafloor, damaging deep-sea ecosystems and devastating the gulf's seafood industry.

Compare the feather-like gills in front of the walking legs of both shrimp. Dark contaminants found on the gills of the specimen collected in 2011 *(bottom)* highlight the tissue damage—which likely adversely affects respiration—found in shrimp and fish after the spill. Scientists also found that the number of different species and the abundance of each species in these ecosystems had declined—but whether the oil spill is the only cause is still unknown.

MAINE

Eastern White Pine Cone

Pinus strobus

Collected in 2025 from Maine, exact location unknown
US-00892887

The state tree of Maine and Michigan, the eastern white pine towers majestically at fifty to eighty feet (15–24 m) at maturity. Preferring cool, humid climates and moist, sandy soil, these trees once ranged across the northeastern and north-central United States and Canada and south through the Appalachian Mountains. Their long, narrow cones are three to six inches (8–15 cm) long and hold winged seeds that are dispersed by wind. Eastern white pines are cultivated for the Christmas tree market and for ornamental gardens, where they attract songbirds and other native wildlife.

MAINE

Northern Lobster

Homarus americanus

Collected in 2024 from Maine, exact location unknown
USNM 1754437

Maine's state crustacean can grow up to forty-four pounds (20 kg) and twenty-five inches (64 cm). That makes the northern lobster the heaviest living arthropod species in the world. They live in cold, shallow waters along the Atlantic coast from northern Canada to North Carolina but are rare south of New Jersey. They are easily recognized by their

two long antennae, which they use to detect odors, and their asymmetrical front claws, which they use to hold, crush, and tear into mussels, clams, snails, starfish, sea urchins, bristle worms, and other marine life on the seafloor. Given their size and aggressiveness, northern lobsters' main predators are codfish, seals, and humans.

MARYLAND

Rock Greenshield Lichens

Flavoparmelia baltimorensis

Collected in 1938 from Montgomery County, Maryland
US-00265123

Even unassuming specimens such as this lichen tell big stories about environmental changes. Lichens are a symbiosis between a fungus and an alga. They absorb nutrients—and pollutants—from the atmosphere. One such pollutant is lead, a known neurotoxin to humans and other animals. Museum scientists compared lead levels in lichens collected before *(pictured)* and after the completion of the American Legion Memorial Bridge in 1962. This bridge carries a high volume of traffic on Interstate 495 over the Potomac River near Plummers Island, where this specimen was collected. Lichens exposed to car exhaust from bridge traffic contained significantly higher levels of lead and exhibited damage at the cellular level. However, lead levels in lichens—and in the atmosphere—dropped after leaded gasoline was banned in the 1970s.

MARYLAND

Blue Crab

Callinectes sapidus

Collected from Chesapeake Bay, Maryland, year unknown
Gift of the Delaware Museum of Nature & Science
USNM 1578938

The blue crab's life cycle takes it all over the Chesapeake Bay, making this species a great indicator for the overall health of our country's largest estuary. Females spawn at the mouth of the bay, releasing larvae into the ocean. Young crabs migrate farther into the bay to shelter amid eelgrass beds. Blue crabs are food for many species, from striped bass and great blue herons to humans. In recent years, overfishing and loss of eelgrass have severely reduced their numbers, affecting both recreational and commercial harvests. Smithsonian scientists track blue crabs to learn how environmental conditions affect their population. Blue crabs are a vital part of the Chesapeake Bay's fishing industry, and this information can help develop sustainable management strategies and preserve one of Maryland's cultural symbols.

Tympanuchus cupido

MASSACHUSETTS

Heath Hen (extinct)

Tympanuchus cupido cupido

Collected in 1902 from Martha's Vineyard, Massachusetts
USNM 462954

One of the first birds Americans tried to preserve via legislation, the heath hen was a heavy-bodied bird that once fed and bred in the scrublands and barrens along the Atlantic coast from southeastern Maine to northern Virginia. Both Native Americans and European settlers hunted the heath hen for food, but human population growth led to overhunting and habitat loss. By 1870, no heath hens lived on the mainland—they were confined to the island of Martha's Vineyard in Massachusetts. Hunting bans and the establishment of a reserve protected the remaining heath hens until a fire swept their breeding grounds in 1916, followed by several harsh winters and failed captive breeding attempts. The last heath hen was seen in 1932.

MICHIGAN

Petoskey Stone

Hexagonaria percarinata

Lived 388–383 million years ago
Donated in 1890, collected from northern Michigan
USNM PAL 23832

The official stone of Michigan, Petoskey stones were formed by glaciers pulling fossilized rugose corals from the bedrock, grinding their surfaces smooth, and depositing them along the shores of Lake Michigan. Rugose (meaning "wrinkled") corals are an extinct type of coral, which were abundant 470 to 252 million years ago. Like many corals today, they lived on the seafloor either individually or in large colonies, and their skeletons were made of calcite. This particular coral species is found only in Michigan's Gravel Point Formation, mostly around the town of Petoskey in the northwest portion of the lower peninsula.

MICHIGAN

Copper

Cu

Acquired in 1972, collected from Keweenaw Peninsula, Michigan
Gift of Arthur Montgomery
USNM 123608

Archeological evidence demonstrates that copper has been mined in northwest Michigan's Keweenaw Peninsula as far back as 5000 BCE to produce tools. In the nineteenth and early twentieth centuries, copper mining became an important industry for the Michigan economy. The copper mined from the Keweenaw Peninsula is the purest copper found anywhere on Earth. That purity means the copper hasn't formed compounds with other elements—but the deposits do contain toxins like mercury and arsenic. The United States is the world's fifth largest copper producer, though most US copper today comes from Arizona. In recent years, there has been renewed interest in Michigan's copper to meet the growing global demand for telecommunications and electronic components, as well as plumbing and construction materials.

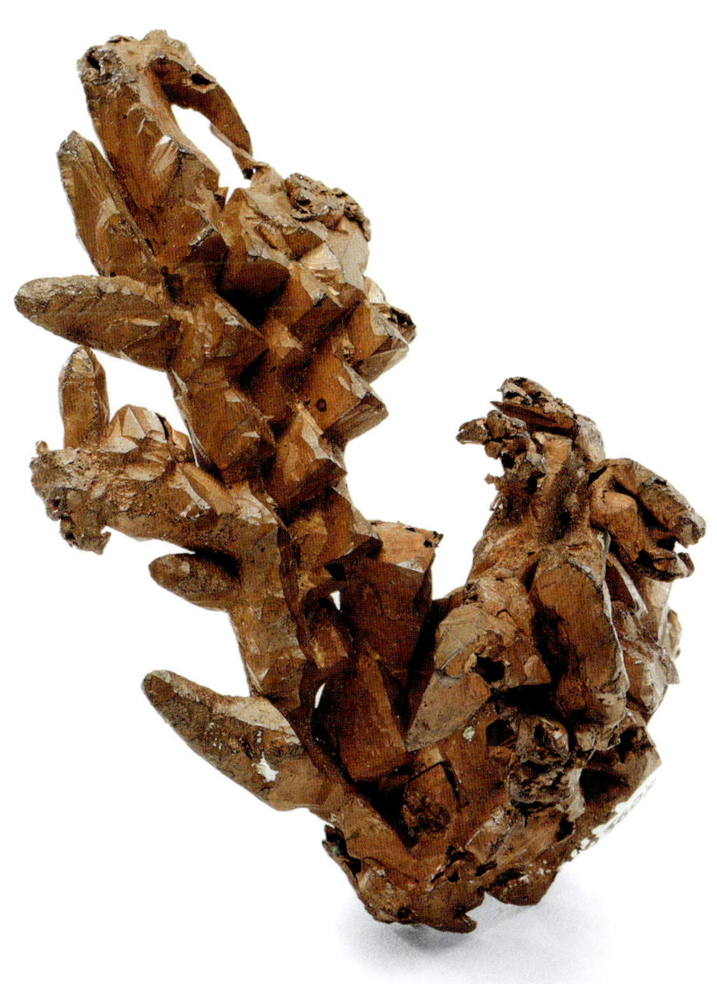

MINNESOTA

Twelve-spotted Skimmer

Libellula pulchella

Collected in 1907 from Ramsey County, Minnesota

USNM ENT 00485157

One of over 3,000 dragonfly species worldwide and named for the dark spots on its wings, the twelve-spotted skimmer is found in all forty-eight contiguous US states. They live in wetland habitats near still water, where they hunt for mosquitoes, midges, and small flies by flying low, or skimming, over the surface of the water before returning to a familiar perch. Male skimmers are often territorial over their feeding and breeding areas, aggressively chasing off competitors and sometimes harassing females. Female skimmers' aerial feats include laying eggs while in flight. They do this by striking the water's surface with the tip of their abdomen.

MINNESOTA

Sea Star

Hudsonaster narrawayi

Lived 457–449 million years ago

Acquired in 1928, collected from Minnesota, exact location unknown

USNM PAL 92631

Sea stars first evolved 480 million years ago when Earth was a lot warmer and shallow seas covered much of the land. Today, they live in every ocean basin, from rocky tidepools to warm, sandy seabeds to frigid mud on the ocean floor. However, you can no longer find them in Minnesota because there are no freshwater sea stars. This and other marine fossils in Minnesota are evidence that portions of this state were once part of a tropical shallow sea south of the equator that covered most of the North American continent.

Chahta i̱ lokka (Choctaw Dress)

Maggie Billie (Choctaw, ca. 1907–2008)
Made in 1932 in Neshoba County, Mississippi
Gift of Frances Theresa Denmore
NMNH E380093 (dress), E380094 (apron)

The diamond and half-diamond patterned bands in this Choctaw dress pay homage to the eastern diamondback rattlesnake from the Choctaw's ancestral homeland in what is now Alabama, Louisiana, and Mississippi. These predators played an important role in Choctaw agriculture by removing rodents from corn fields. Half-diamonds can also represent the hilly Choctaw homelands, literal journeys, and life's highs and lows. The Choctaw were relocated under the Indian Removal Act of 1830. Thousands of people died on the journey to Oklahoma, and those who stayed in Mississippi had to give up their tribal sovereignty until 1945. Today, the Oklahoma, Mississippi, and Louisiana bands of Choctaw maintain cultural ties through shared clothing styles, food traditions, and dances.

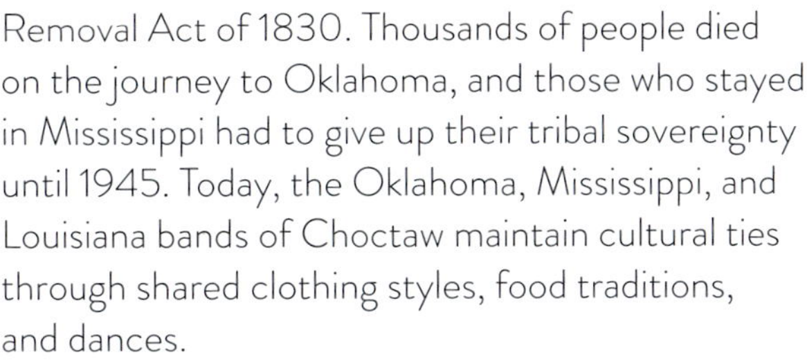

Maggie Billie, shown here in 1933 wearing her regalia, was an accomplished seamstress whose elaborate outfit showcases handsewn detailing and cutouts.

MISSOURI

American Paddlefish (adult skeleton and juvenile in fluid)

Polyodon spathula

Adult *(bottom)* collected from Missouri River, year unknown
Juvenile *(top)* collected in 1992 from Boone County, Missouri
USNM 403936, 397330

The United States is home to the last remaining species in this ancient fish lineage: the American paddlefish, which is among the largest freshwater fishes in North America. It hasn't changed much over its evolutionary history, forming a living bridge between the present and distant past. Its closest relatives are sturgeons, though they diverged 180–100 million years ago. Paddlefish once lived throughout the Great Lakes but are now only found in the Mississippi watershed. Paddlefish use their snout to sense electrical currents, which helps them find microscopic food and avoid obstacles in murky waters.

MONTANA

Plains Bison Calf

Bison bison bison

Collected in 1954 from Moiese, Flathead Indian Reservation, Montana
USNM 300326

The official mammal of the United States, bison once roamed North America's grasslands by the millions. By 1886, though, only a few hundred plains bison (also called buffalo) remained in the United States. Thanks to the efforts of Indigenous communities and conservationists throughout the twentieth century, about 20,000 buffalo are alive in the wild today, and several hundred thousand are kept in commercial herds. This calf came from the former National Bison Range, now the CSKT Bison Range, managed by the Confederated Salish and Kootenai Tribes.

Bison shape the landscape in many ways. They aerate soils, disperse native seeds, and distribute nutrients in their dung as they migrate. They create depressions in the soil that retain rainwater and provide habitat for other animals. Bison are key to healthy grasslands, which store lots of carbon. The loss of bison destabilized ecosystems and devastated Native communities. Today, the growth of wild and managed herds has begun to revive the concept of bison as culturally and ecologically important animals.

NEBRASKA

Tortoise

Stylemys nebrascensis

Lived 34–32 million years ago
Collected prior to 1899 from northwest Nebraska
USNM PAL 358863

Meet *Stylemys*, the first fossil tortoise discovered in North America. This fossil is immediately recognizable as a tortoise and appears to be closely related to modern gopher tortoises, showing that this body plan hasn't changed much in 30 million years. Today, the United States has just one native tortoise genus, *Gopherus*, made up of four species of gopher and desert tortoises. Much like their ancient ancestors, they live in temperate to subtropical environments from southern California to Florida. Gopher tortoises are a keystone species, or one that has a major effect on its ecosystem. In their case, gopher tortoise burrows provide critical habitats for hundreds of other animal species.

NEVADA

Great Basin Bristlecone Pine Cone

Pinus longaeva

Collected in 1891 from Charleston Mountain, Nevada
US-02744559

The official state tree of Nevada, Great Basin bristlecone pines are native to the high, windblown mountains of California, Nevada, and Utah. These knotted, twisted trees grow up to fifty feet (15 m) tall, have cones that are between two and four inches (5–10 cm) long, and can live for thousands of years in a harsh terrain where few other plants thrive. In fact, they are some of the oldest single organisms in the world. The oldest living bristlecone pine is over 4,800 years old!

The tree rings of these long-lived pines provide scientists with data to better understand and more precisely date changes in climate over the last several thousand years. This cone was collected by botanists Frederick Coville and Frederick Funston as part of the US Department of Agriculture's Death Valley expedition in 1891 to study the distribution of plants and animals in the Southwest.

NEVADA

Twinned Gold Crystals

Au

Acquired in 2023, collected from Lander County, Nevada
USNM 177827-1

Gold isn't just for jewelry; it conducts electricity, resists corrosion, and is very malleable, making it an important component of electronics. It's in everything from cell phones and satellite systems to medical devices and green technologies.

Most gold in the United States is mined in Nevada, where igneous activity concentrated it along the edges of tectonic plates approximately 23–5 million years ago. This specimen displays many twinned crystals—that is, multiple crystals sharing atoms in a symmetrical fashion. Look closely for crystals that appear to be built on other crystals but are oriented in a different direction. You will also see tiny crystals of naumannite on the surface of the gold crystals, giving them a tarnished look.

NEW HAMPSHIRE

Fluorite

CaF_2

Acquired in 2019, collected from Cheshire County, New Hampshire
Gift of Robert Lavinsky
USNM 177433

The mineral fluorite naturally occurs in deposits in many states, including New Hampshire. In its pure form, this mineral is colorless and transparent, but it is often found in a wide range of colors including shades of green, purple, and yellow. Copper impurities likely give this specimen its bright green color. Under ultraviolet light, many specimens of this mineral exhibit fluorescence—a phenomenon that got its name from fluorite. Fluorite is not only beautiful; it also has many properties that make it useful in industry. It is used as a flux to increase fluidity and remove impurities in steel and aluminum smelting processes, as a lining material in furnaces and kilns, and as a component in telescope and microscope lenses. Fluorite is also a major component in hydrofluoric acid, which is used to make Prozac, Teflon, industrial etching and cleaning products, and rust inhibitors, just to name a few uses.

NEW JERSEY

Bog Turtle Shell

Glyptemys muhlenbergii

Collected in 1930 from Monmouth County, New Jersey
USNM 288353

New Jersey's official state reptile is also the smallest turtle in North America. Bog turtles reach no more than 4.5 inches (11.4 cm) long and weigh less than four ounces (113 g). Nevertheless, these little guys have surprisingly long lifespans, averaging twenty to thirty years, and they sometimes live up to sixty years in the wild! Bog turtles spend much of that time basking in the sunlight surrounded by slow-moving water and low grasses or buried in mud to regulate their temperatures. They hibernate during the winter under the dense roots of trees and shrubs or under rocks. Bog turtles are listed as threatened or endangered because their habitats have been reduced and fragmented as a result of development.

NEW JERSEY

Luna Moth

Actias luna

Collected in 1956 from Mercer County, New Jersey
USNM ENT 01978324

One of the larger moths in North America, the luna moth has an average wingspan of 4.5 inches (11.4 cm) and can be found in deciduous forests east of the Great Plains, from central Quebec to Florida. Common but rarely seen, they are active at night and blend into the foliage of the trees they live in and feed on hickory, birch, persimmon, sweet gum, walnut, and sumac.

Luna moths have some interesting ways to avoid their predators. As caterpillars, they make a warning clicking noise when attacked and regurgitate a bad-tasting fluid. In their adult form, the long tails on their hindwings confuse bats—who use echolocation to hunt—and fool them into attacking the tails, allowing the moths to survive another day.

NEW MEXICO

Gypsum Sand

$CaSO_4 \cdot 2H_2O$

Collected prior to 1938 from White Sands National Park, New Mexico
NMNH 101918

White Sands National Park is the largest gypsum sand dune field in the world. These sparkling dunes were once an ancient seabed full of the mineral gypsum. Geological uplift 70 million years ago exposed the seabed, then rainwater dissolved the gypsum and flowed to the flat lake bed, where the gypsum was deposited. When the lake dried up 10,000 years ago, strong winds broke the crystals into a fine sand and pushed them into massive dunes. Gypsum becomes less water soluble at higher temperatures, making the arid desert climate of this area of New Mexico perfect for gypsum sand to accumulate.

NEW MEXICO

Temnospondyl Amphibians

Anaschisma browni

Lived 224–215 million years ago
Collected in 1947 from Santa Fe County, New Mexico
USNM V 18495

Anaschisma was a giant freshwater amphibian with small limbs, a large flat head, and a wide mouth filled with sharp teeth, similar to a salamander version of a crocodile-like predator. Adults grew up to nine feet (2.7 m) long.

All the fossils in this bone bed occur in a single rock layer only four inches (10 cm) thick, revealing that they accumulated during a single, rapid event like a flash flood. The long bones also tend to point in the same direction, indicating that they were moved by the water's current. However, the bones are completely jumbled, suggesting that the carcasses had already decayed and come apart before being washed to their final resting place.

NEW YORK

Sea Scorpion

Eurypterus remipes

Lived 423–419 million years ago
Collected prior to 1890 from Herkimer County, New York
USNM PAL 23693

New York's state fossil, *Eurypterus remipes*, represents an extinct group of arthropods (the phylum that includes crustaceans, insects, and spiders, among others) that hunted for soft-bodied invertebrates and scavenged for food in subtropical marine waters. Eurypterids, or sea scorpions, were related to horseshoe crabs, true scorpions, and spiders.

The two sea scorpions in this fossil are seen as if from above. Look closely to spot their two large paddle-like arms that were used for swimming. Other smaller appendages that would have been used to kill and hold prey appear fainter in this fossil. The rounded carapace, or upper section of its exoskeleton, protected the fused head and thorax. Their abdomens were divided into twelve segments with a long, sharp, serrated spine at the end.

NEW YORK

Canada Goose Snarge and Airplane Fragments

Branta canadensis

Collected in 2009 from New York, New York
Not accessioned

On January 15, 2009, US Airways Flight 1549 struck a flock of birds, disabling both engines. Captain Chesley Sullenberger landed the plane in the middle of the Hudson River, and everyone onboard survived. The birds, however, did not, and their remains (or "snarge") came to the museum for identification.

After birds collide with airplanes, what's left isn't pretty, but researchers from the Smithsonian's Feather Identification Lab compare samples from airline accidents with specimens and DNA from the museum's collections to make an identification. Each year, bird strikes kill thousands of birds, put people in danger, and cause millions of dollars in damage to US commercial and military planes. Solutions to this problem, which can range from making airfields less attractive to birds to designing engines that withstand bird weights to reducing the populations of problematic birds, depend on the species involved.

NORTH CAROLINA
Ilex vomitoria Aiton
Location: Fort Fisher State Historical Site, North Carolina. Ocean side of Route 421. About sea level. Sandy soil. Growing in clumps.
Collectors: William L. Merrill, William C. Sturtevant, and Christian Feest.
Date: 11 March 1973
Ilex vomitoria Aiton
Local name: yaupon
Local use: Used locally in the past as a substitute for tea. To prepare, parch leaves and boil in water about same length of time as required to prepare regular tea. Alternate method of preparation: after parching, crumble leaves and then boil as before. If prepared from unparched and green leaves, the resulting tea can be used as a laxative.
UNITED STATES
2669021
NATIONAL HERBARIUM
Image No.
03350731

NORTH CAROLINA

Yaupon

Ilex vomitoria

Collected in 1973 from New Hanover County, North Carolina
US-03350731

Caffeine lovers, rejoice! If you live in the southeastern United States, you are likely surrounded by wild tea in the form of North America's only native caffeinated plant. Many Indigenous communities consumed yaupon for medicinal, spiritual, or social purposes. European colonists enjoyed this tea, and it became a protest beverage against the British during the Revolutionary War. During the Civil War, Americans drank yaupon tea widely due to coffee and tea shortages. Afterward, it suffered from associations with poverty, wartime hardships, and negative stereotypes against African American and Indigenous communities. Recently, there has been renewed commercial interest in this plant as a more sustainable and environmentally friendly alternative to coffee and imported tea.

NORTH CAROLINA

Mega-toothed Shark Teeth

Otodus megalodon (formerly *Carcharocles megalodon*)

Lived 5–3 million years ago
Collected c. 1970s from Beaufort County, North Carolina
USNM PAL 256311, PAL 355870, PAL 355882,
PAL 355894, PAL 356967

North Carolina's state fossil isn't a whole organism; it's just the teeth of the mega-toothed shark commonly known as "megalodon." This extinct species is closely related to mako sharks and grew up to three times the size of a modern-day great white shark. Imagine a mouth the average person could easily step through, lined with several rows of sharp, 6-inch (15.2 cm) teeth, attached to a 60-foot (18 m), 112,000-pound (50,802 kg) shark! The museum's life-size model of this shark is fifty-two feet (15.8 m) long.

Paleontologists have found megalodon bite marks along with fragments and whole teeth embedded in the fossil bones of their prey, such as whales, seals, sea cows, and sea turtles. Most of what we know about this species comes from their teeth, vertebrae, and fossilized feces, as most of their skeletons are made of cartilage, which rarely fossilizes.

NORTH DAKOTA

Ginkgo Leaf

Ginkgo wyomingensis

Lived 57 million years ago
Acquired c. 1980s, collected from Morton County, North Dakota
USNM PAL 799145

This North American ginkgo fossil hearkens back to a time when this group of plants was more widely distributed than it is now. The order Ginkgoales traces its evolutionary origins back to the Permian Period around 270 million years ago, but only one species, *Ginkgo biloba*, remains today. Modern ginkgoes are native to China and the Korean peninsula and are extinct in the wild, yet you'll find them lining city streets around the world, their fan-shaped leaves turning bright yellow each autumn. These distinctive leaves haven't changed much over time, meaning scientists can compare cellular structures between modern and fossil ginkgo leaves, using the density of pores to estimate atmospheric carbon dioxide levels.

NORTH DAKOTA

Eared Grebe

Podiceps nigricollis

Collected in 2012 from Bottineau County, North Dakota
USNM 647594

During the 1872–73 US Boundary Survey, museum scientists collected many natural history specimens. The Smithsonian collected specimens from these areas again 140 years later to document environmental and evolutionary changes. Specimens such as this eared grebe, collected in 2012, were also used in a companion project that studied salt-gland development in water birds.

Eared grebes nest in shallow freshwater wetlands throughout western North America from Mexico to Canada. Monogamous pairs perform elaborate courtship dances that involve running across the water. In early autumn, they flock to salt lakes in Utah and California to fill up on brine shrimp on their way south for winter. Salt glands near their eyes help them expel the extra salt.

NORTHERN MARIANA ISLANDS

Spindle Bomb
Igneous rock
Collected prior to 1955 from Pagan Island, Northern Mariana Islands
USNM 109243-5

During some explosive volcanic eruptions, molten or partially molten lava "bombs" of basalt are ejected high into the air. Some of these lava bombs spin during flight, resulting in an elongated shape with pinched ends, called spindle bombs. This specimen came from Mount Pagan, one of the largest and most active volcanoes in the Northern Mariana Islands, a volcanic arc marking where the Pacific Plate subducts under the Philippine Plate. The fourteen islands and more than sixty seamounts, or underwater mountains, in this chain are all active volcanoes. The largest eruption of Mount Pagan in historical times led to the evacuation of the island's small population in 1981. This eruption continued for four years, with numerous small eruptions since.

OHIO

Passenger Pigeon (extinct)

Ectopistes migratorius

Collected in 1914 from the Cincinnati Zoo and Botanical Garden, Ohio
USNM 223979

In just one hundred years, passenger pigeons went from billions to zero. Their flocks once spread across the deciduous forests of North America east of the Rocky Mountains. Overhunting and deforestation throughout the nineteenth century did what no natural predator could, and by the 1890s, only a few birds remained. The Cincinnati Zoo tried to breed captive pairs to no avail. The last known individual, Martha, died on September 1, 1914, and was shipped to the Smithsonian for preservation. Martha *(pictured)* is one of the museum's most precious specimens, an ambassador urging us to do more to prevent the extinction of other organisms.

OKLAHOMA

Scissor-tailed Flycatcher

Tyrannus forficatus

Collected in 1905 from Grady County, Oklahoma
USNM 574813

Oklahoma's official state bird is known for its long, forked tail that the males use in aerial acrobatics during courtship displays. They range eight to fourteen inches (20–36 cm) in length and weigh between one and two ounces (28–57 g). Scissor-tailed flycatchers are migratory birds that spend April to October in their breeding grounds of Texas, Oklahoma, Arkansas, Kansas, Missouri, and Louisiana. They overwinter in southern Mexico and along the Pacific coast of Central America. These birds prefer to perch on treetops, fences, and utility lines to watch and wait for insect prey, and they aggressively defend their nests.

OKLAHOMA

Mexican Free-tailed Bat

Tadarida brasiliensis

Collected in 1955 from Major County, Oklahoma
USNM 287929

Although it is the official flying mammal of both Oklahoma and Texas, the Mexican (or Brazilian) free-tailed bat is one of the most abundant mammals in North America. It can be found across the southern part of the United States from Florida to California. These small bats are about 3.5 inches (9 cm) long including their tails, which account for about half their length. During the day, they roost in large numbers in caves and under bridges. At night, their large, wide, round ears help them find insect prey via echolocation. Each summer, more than a million bats emerge from Oklahoma's four known maternity caves at dusk for their nightly feast.

OREGON

Oregon Hairy Triton

Fusitriton oregonensis

Collected in 1888 from Lane County, Oregon
USNM MOL 224570

Oregon's state seashell comes from a predatory sea snail that lives in the shallow waters along the northwestern coast of North America from Alaska to California. Oregon hairy tritons grow up to three to five inches (8–13 cm) long, and their shells are covered with bristly hairs growing from their thin skin-like outermost layer. Like many other sea snails, the Oregon hairy triton hatches from an egg and spends the juvenile stage of its life as a free-floating larva before undergoing metamorphosis into its adult form. What is extraordinary about this species is the unusually long duration of its larval stage—up to four and a half years!

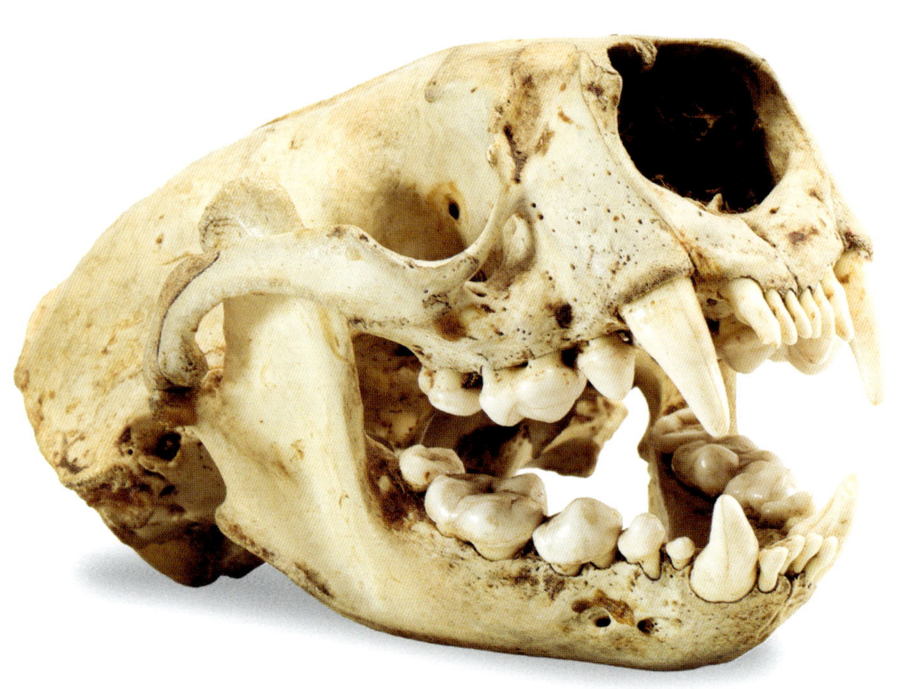

OREGON

Sea Otter Skull

Enhydra lutris

Collected in 1875 from Oregon, exact location unknown

USNM 13460

Sea otters lived up and down the Pacific coast before the fur trade nearly wiped them out completely. This specimen was collected in 1875, when Oregon's sea otters were already extremely rare. Considered a keystone species, sea otters have an outsized role in their ecosystems. By eating grazers such as sea urchins and crabs, these plush predators keep kelp forests and seagrass beds healthy and resilient, which in turn provides vital habitat for many other species, protects coastlines from erosion, and stores carbon, helping offset rising carbon dioxide levels.

Expanding legal protections throughout the twentieth century, along with reintroduction efforts in the 1960s and 70s, allowed sea otter populations to recover and extend their range. However, they have not been successfully reintroduced in Oregon—yet.

PENNSYLVANIA

Hellbender Skull

Cryptobranchus alleganiensis

Collected prior to 1885 from Pennsylvania, exact location unknown
USNM 218486

Pennsylvania's state amphibian may have a fierce name, but North America's largest salamander, the hellbender, just wants to be left alone. These amphibians live solitary lives (except during the autumn mating season) in rocky, swift-moving, cool water streams and rivers through the eastern and central United States. Hellbenders rarely leave the water and have adapted to breathe by drawing oxygen from the water through capillaries in the skin folds along their sides. They still have lungs, but they are used mostly for buoyancy.

The presence of these giant salamanders is a sign that the water is clear, clean, and not polluted. Habitat degradation throughout their range threatens their existence, and they are listed as Vulnerable on the International Union for Conservation of Nature's (IUCN) Red List of Threatened Species.

Smithsonian
United States
National Museum.
Institution.

PENNSYLVANIA

Orchard Oriole

Icterus spurius

Collected in 1840 from Cumberland County, Pennsylvania
USNM A150

This orchard oriole is one of the museum's first specimens. It was collected by Spencer F. Baird before the Smithsonian Institution was established. Baird personally donated more than 3,600 birds to the Smithsonian in 1850. As the Smithsonian's first curator (1850–78) and its second Secretary (1878–87), Baird nurtured the expanding Smithsonian collections. Specimens, such as this one, are prepared and preserved as "study skins," in which the skin, feathers, beak, and lower legs are removed from a dead bird, stuffed with cotton, sewn up, and labeled with associated data.

The orchard oriole is often mistaken for its close relative, the Baltimore oriole (*Icterus galbula*) because of their similar size, coloration, and overlapping ranges—hence its species name *spurius* from the Latin word meaning "false."

PUERTO RICO

Guiros and Scraper

Guiro *(bottom)* collected in 1899 from Caguas, Puerto Rico
Gift of Louise Meyer, NMNH E201483A-0
Guiro and scratcher *(top)* collected in 1956 from Rio Piedras, Puerto Rico
Gift of The Catholic University of America, NMNH E395238-0

Jíbaro music started as a folk music style in Puerto Rico's mountainous countryside, blending Spanish colonial and native Taíno elements. The rhythm comes from the guiro, an Indigenous instrument made with a hollow gourd. The guiro's ridges are scraped with stiff wires or a baton to make a shushing or rasping sound. Originally tied to traditional small-scale farmers, jíbaro music has made its way around the world as a symbol of this island's culture. Today, guiros are made of metal, plastic, or wood, but they still accompany stringed instruments such as the cuatro and guitar.

The instrument's name comes from the Taíno word for the fruit of the *higüero* or calabash tree (*Crescentia cujete*). The Taíno are the Indigenous inhabitants of Puerto Rico and the US Virgin Islands.

PUERTO RICO

Octocoral with Brittle Star

Placogorgia rudis and *Asteroschema oligactes*

Specimen collected in 1933 from San Juan, Puerto Rico
Illustration drawn by Elie Cheverlange
USNM COE 50194

The Smithsonian first explored the deep waters off the coast of Puerto Rico in 1933. The expedition was supported by US businessman and engineer Eldridge Johnson (1867–1945), who offered the use of his yacht, outfitted at his expense with state-of-the-art oceanographic equipment including an echo-sounding device provided by the US Navy. Led by Dr. Paul Bartsch, curator of mollusks from 1914–46, the expedition collected many deep-sea species, like this brittle star entwined with a piece of octocoral, and conducted meteorological, physical, and chemical studies of the ocean just north of Puerto Rico. The scientific illustration of the specimen highlights the brittle star to make it easier to see on the octocoral.

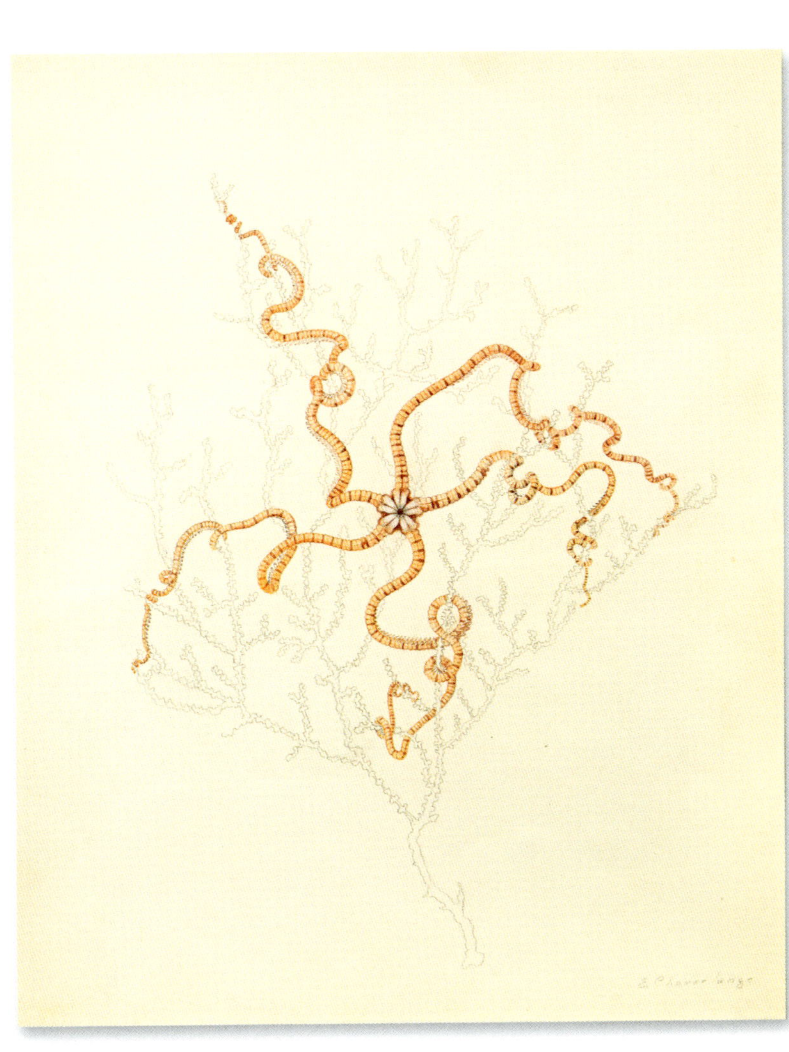

RHODE ISLAND

Longfin Squid

Doryteuthis pealeii

Collected in 1874 from Narragansett Bay, Rhode Island
USNM MOL 814173

Sustainably managed and harvested in US waters, the longfin squid is known for its mild, sweet meat, making it the most common species found as "calamari" on restaurant menus. These marine mollusks do not live very long—less than a year—but they propel themselves quickly through the water by sucking in water and pushing it out forcibly to quickly escape predators. They also have special pigment in their skin cells that change colors and patterns for camouflage.

Reaching lengths of eleven to nineteen inches (28–48 cm), not including their tentacles, longfin squid live over sandy and muddy areas of the continental shelf and upper continental slope from Newfoundland to the Gulf of Venezuela.

COLLECTOR, WALTER A. ANGELL, CENTREDALE, R. I.
NO.
NAME.
SEX.
DATE.
LOCALITY:
5-17-94
608
May 17, 1894

RHODE ISLAND

Scarlet Tanager

Piranga olivacea

Collected in 1894 from Providence County, Rhode Island

USNM 422827

During their summer breeding season in the northeastern, midwestern, and Appalachian United States, male scarlet tanagers really deserve their name. Their feathers are a striking bright red, which contrast with their dark black wings and tail feathers. However, come the fall when they migrate back to their winter grounds in the forest foothills of the Andes Mountains in South America, their red plumage turns a dull yellow and they look remarkably like their female counterparts. Whether standing out or blending in, these birds are not often seen as they keep to the highest branches of the deciduous trees in their range.

SOUTH CAROLINA

Lettered Olive

Oliva sayana

Collected in 1938 from Charleston County, South Carolina
USNM MOL 818835

The state shell of South Carolina, the lettered olive is a predatory sea snail that lives along the Atlantic coast from North Carolina to Florida and along the gulf coast from Florida to Texas. A relatively fast mover along the sandy bottom of shallow and intertidal waters, this sea snail captures bivalves and small crustaceans and drags them under the sand's surface to eat. Like many other sea snails, lettered olives create a whorled shell, usually about 2.25 inches (6 cm) long, to protect their soft bodies. The intricate markings resemble smudged handwriting or hieroglyphs, giving this species its name.

SOUTH CAROLINA

Bumelia Borer

Plinthocoelium suaveolens

Collected from South Carolina, year and exact location unknown
USNM ENT 01978933

The larvae—or juveniles—of bumelia borers tunnel through gum bumelia trees (*Sideroxylon lanuginosum*), feasting on the wood and roots. Adults are also found on this host plant, often feeding on nectar or rotting fruit. These borers are native to the southern United States and are part of the longhorn beetle family, named for their antennae, which are longer than their bodies. The bumelia borer can be identified by its iridescent, metallic green (and sometimes blue) wing coverings and bright orange legs, making it a favorite of bug collectors and photographers.

SOUTH DAKOTA

Iridescent Ammonites

Discoscaphites conradi (x2), *Hoploscaphites nicolleti* (x2), *Sphenodiscus lenticularis*

Lived 72–69 million years ago
Collected in 1982 from Corson County, South Dakota
USNM PAL 376969, PAL 376970, PAL 3276971

The lands we know as the United States were not always *land*. Ammonites—shelled relatives of the modern octopus and squid—swam in the vast Western Interior Seaway that once extended from Alaska to Mexico, covering much of what are now the Great Plains. These tentacled predators died out with the dinosaurs, but the layers of the mineral aragonite in the inner layers of their shells maintain their shimmering iridescent colors that do not degrade over time. You may notice there are three different species of ammonites in this one rock. The Cretaceous seas were filled with a variety of life!

TENNESSEE

Freshwater Cultured Pearls

Megalonaias nervosa

Collected in 1994 from Tennessee, exact location unknown
USNM MOL 854705

Before the 1980s, all freshwater pearls from the United States were natural, but thanks to the efforts of John (1925–2000) and Chessy Latendresse (b. 1943), who started the first freshwater pearl farm in the country, the art of culturing freshwater pearls is alive in Tennessee.

Some freshwater mussels, like the washboard mussel (*Megalonaias nervosa*), can make pearls (like the ones depicted) to protect their soft tissue from irritants such as sand grains. To create cultured pearls, a small piece of mantle—the organ that creates a mussel's shell—is implanted in the mantle of another mussel. The host mussel coats this introduced irritant in layers of aragonite, a mineral with an iridescent sheen. Thus, Tennessee's state gem is created!

TEXAS

Sulfur Crystals in Celestine

S and $SrSO_4$

Collected from Texas, year and exact location unknown
NMNH 115015 00

Texas has many sulfur deposits, and its salt domes are sites for both oil and sulfur operations. The process of sulfur mining often melts sulfur crystals to force the liquid out of the rock, but drilling operations sometimes capture intact sulfur crystals. Sulfur isn't just pretty; its industrial uses make it one of Texas's most valuable minerals. Mostly used to make sulfuric acid for oil refining, wastewater treatment, and fertilizer, sulfur is also added to rubber to make it stronger and more elastic and to insecticides as a deterrent for a range of pests.

TEXAS

Alligator Gar Skull

Atractosteus spatula

Collected prior to 1890 from Kinney County, Texas
USNM 42471

The United States is home to an ancient fish lineage of which only seven species survive today. Alligator gars are the largest members of this family, often reaching a whopping eight feet (2.4 m) long, making them some of the largest freshwater fishes in North America. They live in fresh and brackish water and can gulp air when water oxygen levels get low.

Gars haven't changed much over their evolutionary history, forming a living bridge between the present and distant past. Fossils show they have sported their distinctive "gar-mor" of hard, interlocking diamond-shaped protective scales for millions of years.

US VIRGIN ISLANDS

Ginger Thomas

Tecoma stans

Collected in 1897 from St. Croix, US Virgin Islands
US-00762702

Known as ginger Thomas, fever bush, yellow bells, and yellow elder, the US Virgin Islands' official flower is a fast-growing, drought-resistant shrub that attracts bees, butterflies, and hummingbirds with its fragrant yellow trumpets. This shrub can be found in many different ecosystems from mountain forests to dry scrubland and sandy beaches throughout the southern United States, Mexico, Central America, and northern parts of South America. In the US Virgin Islands, its leaves and flowers are valued for their sunny appearance and uses in natural remedies. Outside its native range, however, the ginger Thomas grows and spreads quickly and is considered invasive.

UNITED STATES NATIONAL HERBARIUM
425238
UNITED STATES NATIONAL HERBARIUM
PURCHASED BY THE U. S. NATIONAL MUSEUM
FLORA OF THE DANISH WEST INDIES,
Island of St. Croix.
Tecoma stans, Juss.
Midland. Jan. 13. 1897.
Collected by Mrs. Rev. J. J. Ricksecker.
No. 21.
Named and distributed by Alfred E. Ricksecker.
00762702

Smithsonian
United States 355384 National Museum. ♂
Anna's Hope, St. Croix
S. T. Danforth
Institution.

US VIRGIN ISLANDS

Bananaquit

Coereba flaveola

Collected in 1926 from St. Croix, US Virgin Islands
USNM 355584

Bananaquits love nothing more than sugar. They sip sweet nectar from flowers with their slender, curved bills, which they also use to pierce flowers at the base to let the syrupy liquid flow. These active foragers will go so far as to steal sugar off tables, hence the Dutch name *suikerdiefje* or "sugar thief." Given that sugar production was an important part of the US Virgin Islands (formerly the Danish West Indies) for more than two centuries, it makes sense that the bananaquit would be designated as their official bird! This highly vocal bird is a critical pollinator found throughout northern South America and on every island of the Caribbean except Cuba.

UTAH

Gila Monster

Heloderma suspectum

Collection date and location unknown
USNM Lab Herp 1808

Utah's state reptile is the United States's largest native lizard, growing up to twenty-two inches (56 cm) long. It's also the only native venomous lizard. Unlike some venomous snakes, which inject venom through hollow fangs, these lizards bite down and chew, which draws the venom up through the grooves in their lower teeth and into their victims. Their black and orange color pattern warns away potential predators.

Gila monsters are solitary desert dwellers that live in rocky regions, scrub, and grasslands up to 5,000 feet (1,524 m) in elevation. They are most active on early spring mornings in parts of Arizona, California, New Mexico, Nevada, and Utah.

VERMONT

Marble

Metamorphic rock

Collected from Chittenden County, Vermont, year unknown
NMNH 17495

It's no wonder that marble, in all its varied hues, is Vermont's state rock. Vermont's quarries, which include the world's largest underground marble quarry, supply stone for everything from kitchen countertops to national monuments. Marble is a metamorphic stone formed when intense heat and pressure recrystallize limestone, a sedimentary rock made from the bodies of tiny marine organisms. This marble's original limestone dated back to the Cambrian and Ordovician periods (about 538–443 million years ago) when Vermont was a shallow marine environment. This marble's pink areas and white veins are calcite deposited by water that flowed through the rock.

Herbarium of North Carolina State University
CASTANEA DENTATA (Marsh.) Borkh.
George P. Johnson
March 1985
PLANTS of VIRGINIA
Shenandoah National Park
Castanea dentata (Marsh.) Borkh. American chestnut
Locality Beahm's Gap Fire Road
Occurrence Rare except as sprouts, great fallen trunks nearby in this second growth forest on old fields, fairly level ground.
Date June 4, 1967
Alt. 600 m
Coll. F. R. Fosberg
No. 48508
Remarks Tree 10-11 m. tall, 17 cm dbh. attacked and nearly girdled by blight, foliage sparse, no fruit set this year; old burs from ground.
UNITED STATES
2669918
NATIONAL HERBARIUM
Image No.
01103548

VIRGINIA

American Chestnut

Castanea dentata

Collected in 1967 from Shenandoah National Park, Virginia
US-01103548

American chestnut trees once grew across the United States from New England to the Appalachian Mountains to the Ohio Valley, with trunks growing over one hundred feet (30 m) tall and ten feet (3 m) in diameter. They produce an edible nut protected by a spiny covering. They were a major source of wood and food for both humans and wildlife until chestnut blight fungus (*Cryphonectria parasitica*) was discovered in New York in 1904 after accidentally being introduced from Asia. By 1950, most of these towering trees had disappeared, although the root systems of some individuals can persist and produce shoots for many years. Today, scientists are trying to breed blight-resistant chestnuts.

VIRGINIA

Great Blue Heron

Ardea herodias

Collected in 2000 from Loudoun County, Virginia
USNM 64902

Great blue herons live in every state except Hawaiʻi. Though they're solitary foragers, they nest in large colonies of several hundred pairs during breeding season, building nests on the ground, in bushes, and even in trees over one hundred feet (30 m) off the ground. Juveniles such as this one have browner plumage than adults, which have distinct black caps and white face markings, as well as shaggy head, neck, and wing feathers. These stately birds are often seen standing still for long stretches of time along shorelines, streams, and estuaries. However, once they spot a fish, frog, or crustacean, they'll strike incredibly quickly, spear it, and swallow it whole.

WASHINGTON

Tacoma Pocket Gopher (extinct)

Thomomys mazama tacomensis

Collected in 1918 from Pierce County, Washington
USNM 231091

This pocket gopher was a subspecies of the Mazama pocket gopher that had a small habitat range in Puget Sound and the Olympic Peninsula. When that land was developed in the mid-twentieth century into suburbs and a gravel mine, these gophers died out. They were last sighted in 1970 and declared extinct in 2013. The loss of a subspecies means its genetic diversity is no longer available, which can make the wider population less able to cope with environmental stressors.

Pocket gophers are the only truly subterranean rodents in North America, spending most of their lives underground. They're called pocket gophers because they have fur-lined cheek pouches they use to transport their food (mainly grasses, forbs, and roots).

WEST VIRGINIA

Spotted Salamander

Ambystoma maculatum

Collected in 1992 from Hampshire County, West Virginia
USNM 596142

Growing up to nine inches (23 cm) long, spotted salamanders are a medium-sized species of mole salamanders, which live only in North America. As the name implies, mole salamanders live in underground burrows. Spotted salamanders can live for decades in the mature deciduous forests of the eastern United States. They spend most of their lives underground and are generally only seen near seasonal pools where they gather to breed. Spotted salamanders have a unique symbiotic relationship with a green alga, which grows inside the egg and developing embryo. This is the only known algal symbiosis involving a vertebrate. Exactly how and why this happens is still a scientific mystery.

WISCONSIN

Maeqtekōs (Dugout Canoe)

Made c. 1775, collected in 1893 from Menominee County, Wisconsin
NMNH E307246

Made from an enormous white pine tree (*Pinus strobus*), this dugout canoe speaks to the Menominee people's forestry expertise. Dugout canoes, some dating back nearly 5,000 years, were made from a variety of tree species using both fire and shell or stone tools to hollow out the trunk.

For thousands of years, the Menominee have managed forest resources to meet their economic, environmental, and social/cultural needs. In 1848,

Chief Oshkosh successfully petitioned the US government for the tribe to stay in part of their ancestral territory. He counseled his people to harvest trees in cycles, a practice that continues today. Though some timber uses have changed, their "forest-first" philosophy hasn't. Harvesting strategies that mimic natural processes keep the forest ecologically healthy and economically profitable, making the Menominee reservation a model for sustainable forestry.

WISCONSIN

Rusty Patched Bumblebee

Bombus affinis

Collected in 1963–65 from Milwaukee, Racine, and Walworth Counties, Wisconsin
USNM ENT 01054162, ENT 01065334, ENT 01064121, ENT 01054876, ENT 01054376, ENT 01004529

The rusty patched bumblebee once lived across the eastern United States and upper Midwest. However, an introduced pathogen—most likely *Nosema bombi*—and exposure to pesticides reduced its population so dramatically that in 2017 it was declared endangered. They are now found mainly in small pockets of Minnesota, Iowa, Illinois, and Wisconsin. Protecting the bees' habitats and planting native pollinator plants can help reduce environmental stressors—and protect other native bumblebee species, which are important pollinators of both crops and wild flowering plants. Certain plants, including tomatoes and cranberries, need bumblebees for pollination, because only these bulky bees can "buzz" to dislodge the pollen.

WYOMING

Geodized Crocodilian Egg

Crocodylia indet.
Lived 50–46 million years ago
Collected in 1930 from Uinta County, Wyoming
USNM V 12597

Though today its winters are cold and windy, millions of years ago Wyoming had a subtropical climate with warm temperatures and large lakes that were perfect for crocodiles. Wyoming's Bridger Formation has plenty of fossil crocodilian bones but very few fossil croc eggshells. Eggs rarely fossilize, mostly because their shells are extremely thin. This one is only 0.03 inches (0.7 mm) thick. Microscopic structures in this eggshell confirm it was laid by a crocodilian, but it never hatched. Instead, mineral-rich water seeped into the egg and deposited sediment. Later, silica crystals formed, creating a sparkling quartz geode.

WYOMING

Freshwater Herring

Knightia eocaena

Lived 50–48 million years ago
Acquired in 1998, collected from Sweetwater County, Wyoming
USNM PAL 482296

Along with the crocodilians, schools of freshwater herring swam in ancient Wyoming's lakes 50 million years ago. Rapid changes in temperature or lake chemistry periodically caused mass die-offs. When the fish sank to the lake bottom, they were buried in sediment before scavengers could get them. The lack of oxygen slowed their decay, and the calm waters kept their skeletons intact.

These mass mortality events happened so often that the Green River Formation is full of layers of *Knightia* skeletons, and they may be the most abundant complete vertebrate fossil in the world. That's why *Knightia* is Wyoming's state fossil (not to be confused with *Triceratops*, its state dinosaur).

ACKNOWLEDGMENTS

A number of colleagues at the National Museum of Natural History were helpful when putting together this book and the accompanying exhibition. We would like to thank:

Michael Ackerson, Kathryn Ahlfeld, Benjamin Andrews, Carrie Beauchamp, Adam Behlke, Kay Behrensmeyer, Joshua Bell, Kate Bemis, Dawn Biddison, Erin Bilyeu, Diana Boudreau, Sarah Bradley, Allison Butler, Matt Carrano, Junko Chinen, Lisa Comer, Evan Cooney, Cari Corrigan, Karolyn Darrow, Megan Dattoria, Kevin de Queiroz, Torsten Dikow, Laura Donnelly-Smith, Nicholas Drew, Stewart Edie, Ericka Gardner, Matt Girard, Gary Graves, Leslie Hale, Ian Herbst, Eric Hollinger, Julie Hoskins, Teresa Hsu, Gwyneira Isaac, Jill Johnson, Kirk Johnson, Ben Kligman, Erin Kolski, Amber Kreiensieck, Igor Krupnik, Michael Lawrence, Dorothy Lippert, Holly Little, Stephen Loring, Darrin Lunde, Kathryn Mathews, Charyn Micheli, Chris Milensky, Matthew Miller, Amanda Millhouse, Jessica Nakano, Martha Nizinski, Juliana Olsson, Lynne Parenti, Myria Perez, John Pfeiffer, Michelle Pinsdorf, Diane Pitassy, Paul Pohwat, Nicholas Pyenson, Kristen Quarles, Andrea Quattrini, Karen Reed, Angela Roberts Reeder, Torbin Rick, Esther Rimer, Amanda Robinson, Ned Rose, Hans Sues, Megan Viera, Warren Wagoner, Shannon Willis, Scott Wing, Jon Wingrath, Ken Wurdack, and Addison Wynn.

SPECIMEN IDS

Page 10 | Top row, starting on the left

1. **Fishfly** (*Dysmicohermes disjunctus*), Washington, USNM ENT 00953295
2. **Sheep moth** (*Hemileuca eglanterina*), Idaho, USNM ENT 01978956
3. **Band-winged grasshopper** (Oedipodinae indet.), Montana, USNM ENT 01978205
4. **Dark red underwing** (*Catocala ultronia*), Montana, USNM ENT 01978253
5. **Melissa blue** (*Plebejus melissa*), North Dakota, USNM ENT 01978717
6. **Nuttall's blister beetle** (*Lytta nuttalli*), North Dakota, USNM ENT 01978480

Second row, starting on the left

7. **Banded alder borer** (*Rosalia funebris*), Washington, USNM ENT 01978337
8. **Snowberry clearwing** (*Hemaris diffinis*), Washington, USNM ENT 01978608
9. **Zerene fritillary** (*Speyeria zerene*), Washington, USNM ENT 02024045
10. **Sculptured pine borer** (*Chalcophora angulicollis*), Idaho, USNM ENT 01978452
11. **Carlinian snapper** (*Circotettix carlinianus*), Idaho, USNM ENT 01978874
12. **Long-horned beetle** (*Xylotrechus longitarsis*), Montana, USNM ENT 01978404
13. **Haldeman's shieldback** (*Pediodectes haldemanii*), South Dakota, USNM ENT 01978983
14. **Nevada buck moth** (*Hemileuca nevadensis*), South Dakota, USNM ENT 01978462

Third row, starting on the left

15. **Pandora pine moth** (*Coloradia pandora*), Oregon, USNM ENT 01978214
16. **Sara orangetip** (*Anthocharis sara*), Oregon, USNM ENT 01978332
17. **Horse fly** (*Hybomitra californica*), Washington, USNM ENT 01978300
18. **Cicada** (*Neoplatypedia constricta*), Idaho, USNM ENT 01978797
19. **Long-legged anabrus** (*Anabrus longipes*), Idaho, USNM ENT 01978280
20. **Caddisfly** (*Hesperophylax designatus*), South Dakota, USNM ENT 01978775
21. **Vashti sphinx** (*Sphinx vashti*), South Dakota, USNM ENT 01978788

Fourth row, starting on the left

22. **Mormon cricket** (*Anabrus simplex*), Oregon, USNM ENT 01978135
23. **Golden jewel beetle** (*Buprestis aurulenta*), Oregon, USNM ENT 01978991
24. **Crackling forest grasshopper** (*Trimerotropis verruculata*), Oregon, USNM ENT 01978221
25. **Buckmoth** (*Hemileuca hera*), Utah, USNM ENT 01978612
26. **Common water strider** (*Aquarius remigis*), Utah, USNM ENT 01978630
27. **Longhorn beetle** (*Crossidius coralinus*), Utah, USNM ENT 01978434
28. **Garden tiger moth** (*Arctia caja*), Utah, USNM ENT 02024048
29. **Rocky Mountain parnassian** (*Parnassius smintheus*), Wyoming, USNM ENT 01978935
30. **Coral-winged grasshopper** (*Pardalophora apiculata*), Wyoming, USNM ENT 01978687
31. **Hide beetle** (*Trox* sp.), South Dakota, USNM ENT 01978727

Fifth row, starting on the left

32. **Anise swallowtail** (*Papilio zelicaon*), Nevada, USNM ENT 01978507
33. **Round-necked longhorn beetle** (*Megacyllene antennata*), Nevada, USNM ENT 01978604
34. **Broad-nosed weevil** (*Ophryastes argentatus*), Utah, USNM ENT 02024010
35. **Large marble** (*Euchloe ausonides*), Utah, USNM ENT 01978876
36. **Botfly** (*Cuterebra lepusculi*), Utah, USNM ENT 01978165
37. **Pine white** (*Neophasia menapia*), Colorado, USNM ENT 02024039
38. **Blue copper** (*Lycaena heteronea*), Colorado, USNM ENT 02024036
39. **Greasewood moth** (*Agapema galbina*), Colorado, USNM ENT 01978597
40. **Fire-necked longhorn beetle** (*Batyle ignicollis*), Colorado, USNM ENT 01978669
41. **Blue pleasing fungus beetle** (*Cypherotylus californicus*), Colorado, USNM ENT 01978804
42. **Melissa blue** (*Plebejus melissa*), Colorado, USNM ENT 02024055
43. **Tachinid fly** (*Adejeania vexatrix*), Colorado, USNM ENT 01978826
44. **Giant lion fly** (*Promachus giganteus*), Colorado, USNM ENT 01978864

Sixth row, starting on the left

45. **Flame skimmer** (*Libellula saturata*), Arizona, USNM ENT 00485306
46. **Giant whip scorpion** (*Mastigoproctus giganteus*), Arizona, USNM ENT 01978163
47. **Owlfly** (*Ascaloptynx appendiculata*), Arizona, USNM ENT 01978640
48. **Glorious scarab** (*Chrysina gloriosa*), Arizona, USNM ENT 01978927
49. **Western horse lubber grasshopper** (*Taeniopoda eques*), Arizona, USNM ENT 01978829

Seventh row, starting on the left

50. **Great purple hairstreak** (*Atlides halesus*), Arizona, USNM ENT 02024034
51. **Flame skimmer** (*Libellula saturata*), Arizona, USNM ENT 00485290
52. **Cactus weevil** (*Cactophagus spinolae*), Arizona, USNM ENT 02024012
53. **California root borer** (*Prionus californicus*), Arizona, USNM ENT 01978426
54. **Western Hercules beetle, female** (*Dynastes grantii*), Arizona, USNM ENT 00990023
55. **Western Hercules beetle, male** (*Dynastes grantii*), Arizona, USNM ENT 00990028
56. **Prairie walkingstick** (*Diapheromera velii*), New Mexico, USNM ENT 01978873
57. **Jewel beetle** (*Lampetis drummondi*), New Mexico, USNM ENT 01978626
58. **Robust toad lubber** (*Phrynotettix robustus*), New Mexico, USNM ENT 01978176
59. **Jewel beetle** (*Sphaerobothris ulkei*), New Mexico, USNM ENT 01978173
60. **Ten-lined June beetle** (*Polyphylla decemlineata*), New Mexico, USNM ENT 01978212
61. **Cactus longhorn beetle** (*Moneilema armatum*), New Mexico, USNM ENT 01978589

Page 11 | Top row, starting on the left

1. **Imperial moth** (*Eacles imperialis*), Virginia, USNM ENT 01978119
2. **Long-tailed giant ichneumonid wasp** (*Megarhyssa macrurus*), Virginia, USNM ENT 01978730
3. **Regal moth** (*Citheronia regalis*), Virginia, USNM ENT 01978116
4. **Dobsonfly** (*Corydalus cornutus*), North Carolina, USNM ENT 01174844
5. **Scorpionfly** (*Panorpa lugubris*), North Carolina, USNM ENT 01978342
6. **Rove beetle** (*Platydracus maculosus*), North Carolina, USNM ENT 01978494

Second row, starting on the left

7. **Tile-horned prionus** (*Prionus imbricornis*), Virginia, USNM ENT 01978842
8. **Dragonhunter** (*Hagenius brevistylus*), Virginia, USNM ENT 00482964
9. **Eastern Hercules beetle** (*Dynastes tityus*), Virginia, USNM ENT 02024016
10. **Wheel bug** (*Arilus cristatus*), Virginia, USNM ENT 01978326
11. **Bumelia borer** (*Plinthocoelium suaveolens*), South Carolina, USNM ENT 01978933
12. **Jewel beetle** (*Buprestis fasciata*), South Carolina, USNM ENT 01978443
13. **Common buckeye** (*Junonia coenia*), South Carolina, USNM ENT 01978263
14. **Sleepy orange** (*Eurema nicippe*), South Carolina, USNM ENT 01978192
15. **Luna moth** (*Actias luna*), South Carolina, USNM ENT 01978210

Third row, under Alabama

16. **Laurel swallowtail** (*Papilio palamedes*), Alabama, USNM ENT 01978962

Fourth row, starting on the left

17. **Bald laphria** (*Laphria apila*), Alabama, USNM ENT 01978295
18. **Live-oak root borer** (*Archodontes melanopus*), Alabama, USNM ENT 01978279
19. **Sweetbay silkmoth** (*Callosamia securifera*), Alabama, USNM ENT 01978316
20. **Broad-winged katydid** (*Microcentrum rhombifolium*), Florida, USNM ENT 01978171
21. **Palmetto conehead** (*Belocephalus sabalis*), Florida, USNM ENT 01978137
22. **Straight-nosed weevil** (*Brentus anchorago*), Florida, USNM ENT 02024014
23. **Florida bee killer** (*Mallophora bomboides*), Florida, USNM ENT 01978938
24. **Large orange sulphur** (*Phoebis agarithe*), Florida, USNM ENT 01978900
25. **Eastern lubber grasshopper** (*Romalea microptera*), Florida, USNM ENT 01978121
26. **Spined green stink bug** (*Loxa flavicollis*), Florida, USNM ENT 01978353
27. **Grizzled mantis** (*Gonatista grisea*), Florida, USNM ENT 01978172
28. **Palmetto weevil** (*Rhynchophorus cruentatus*), Florida, USNM ENT 02024015
29. **Atala butterfly** (*Eumaeus atala*), Florida, USNM ENT 02024037

Fifth row, starting on the left

30. **Giant water bug** (Belostomatidae indet.), Georgia, USNM ENT 01978161
31. **Red velvet ant** (*Dasymutilla occidentalis*), Georgia, USNM ENT 01978674
32. **Great purple hairstreak** (*Atlides halesus*), Georgia, USNM ENT 02024033

Sixth row, starting on the left

33. **Broad-winged katydid** (*Microcentrum rhombifolium*), Georgia, USNM ENT 01978798
34. **Eastern eyed click beetle** (*Alaus oculatus*), Georgia, USNM ENT 01978768
35. **Zebra longwing** (*Heliconius charithonia*), US Virgin Islands, USNM ENT 01978582
36. **Robber fly** (*Efferia stylata*), US Virgin Islands, USNM ENT 01978701
37. **Ornate bella moth** (*Utetheisa ornatrix*), US Virgin Islands, USNM ENT 01978345
38. **Tiger moth** (*Composia credula*), US Virgin Islands, USNM ENT 01978146
39. **Jewel beetle** (*Polycesta porcata*), US Virgin Islands, USNM ENT 01978764
40. **Longhorn beetle** (*Eburia quadrimaculata*), Puerto Rico, USNM ENT 01978376
41. **Longhorn beetle** (*Chlorida festiva*), Puerto Rico, USNM ENT 01978944
42. **Titan sphinx** (*Aellopos titan*), Puerto Rico, USNM ENT 01978889
43. **Longhorn beetle** (*Stenodontes exsertus*), Puerto Rico, USNM ENT 01978742

Page 12 | clockwise from top

1. **Western meadowlark** (*Sturnella neglecta*), Nebraska, USNM 155878
2. **Mountain bluebird** (*Sialia currucoides*), Idaho, USNM 239446
3. **Bananaquit** (*Coereba flaveola*), US Virgin Islands, USNM 355584
4. **Northern cardinal** (*Cardinalis cardinalis*), Georgia, USNM 168916
5. **Cerulean warbler** (*Setophaga cerulea*), Indiana, USNM 118436
6. **Carolina parakeet** (*Conuropsis carolinensis*), Florida, USNM 220629
7. **American goldfinch** (*Spinus tristis*), Minnesota, USNM 221640
8. **Purple martin** (*Progne subis*), Texas, USNM 221718
9. **Painted bunting** (*Passerina ciris*), Texas, USNM 153465

Page 14

1. **Johnstone's junonia** (*Scaphella junonia johnstoneae*), Alabama, USNM MOL 598100
2. **Black abalone** (*Haliotis cracherodii*), California, USNM MOL 3406
3. **Channeled whelk** (*Busycotypus canaliculatus*), Delaware, USNM MOL 806855
4. **Horse conch** (*Triplofusus giganteus*), Florida, USNM MOL 124855
5. **Knobbed whelk** (*Busycon carica*), Georgia, USNM MOL 535350
6. **New England neptune** (*Neptunea lyrata decemcostata*), Massachusetts, USNM MOL 708793
7. **Bay scallops** (*Argopecten irradians*), New York, USNM MOL 516253
8. **Scotch bonnet** (*Semicassis granulata*), North Carolina, USNM MOL 467406
9. **Oregon hairy triton** (*Fusitriton oregonensis*), Oregon, USNM MOL 224570
10. **Hardshell clam/Quahog** (*Mercenaria mercenaria*), Rhode Island, USNM MOL 34432
11. **Lettered olive** (*Oliva sayana*), South Carolina, USNM MOL 818835
12. **Lightning whelk** (*Sinistrofulgur perversum*), Texas, USNM MOL 367945
13. **Eastern oyster** (*Crassostrea virginica*), Virginia, USNM MOL 98050
14. **Olympia oyster*** (*Ostrea lurida*), Washington, USNM IZ 1548589

* *This native oyster was grown in Similk Bay on Swinomish Tribal Lands by the Swinomish Shellfish Company. Gift of Salinity Seafood and More.*

Page 16

1. **Gray pine** (*Pinus sabiniana*), California, US-03465025
2. **Torrey pine** (*Pinus torreyana* var. *insularis*), California, US-02745202
3. **Pond pine** (*Pinus serotina*) opened, South Carolina, US-03465023
4. **Pond pine** (*Pinus serotina*) unopened, South Carolina, US-02745145
5. **Single-leaf pinyon pine** (*Pinus monophylla*), California, US-02744930
6. **Pitch pine** (*Pinus rigida*), New Jersey, US-02745117
7. **Eastern white pine** (*Pinus strobus*), Maine, US-00892887
8. **Sugar pine** (*Pinus lambertiana*), California, US-00892888
9. **Coulter pine** (*Pinus coulteri*), California, US-02744698
10. **Southwestern white pine** (*Pinus strobiformis*), Arizona, US-02745162
11. **Great Basin bristlecone pine** (*Pinus longaeva*), Nevada, US-02744559
12. **Shortleaf pine** (*Pinus echinata*), Mississippi, US-02744716
13. **Red pine** (*Pinus resinosa*), New York, US-02745098
14. **Longleaf pine** (*Pinus palustris*), Florida, US02744986

Page 68

1. **Cicada killer with dog-day cicada** (*Sphecius speciosus* with *Neotibicen canicularis*), USNM ENT 01978548

2. Broad-winged katydid (*Microcentrum rhombifolium*), USNM ENT 01978451
3. Red-spotted purple (*Limenitis arthemis*), USNM ENT 01326590
4. Dobsonfly (*Corydalus cornutus*), USNM ENT 00952846
5. Dog-day cicada (*Neotibicen canicularis*), USNM ENT 02024099
6. Brown prionid (*Orthosoma brunneum*), USNM ENT 01474371
7. Red-legged buprestis (*Buprestis rufipes*), USNM ENT 01939277
8. American bumblebee (*Bombus pensylvanicus*), USNM ENT 01039362
9. Gold-necked carrion beetle (*Nicrophorus tomentosus*), USNM ENT 00833644
10. Purple tiger beetle (*Cicindela purpurea*), USNM ENT 02024023
11. Cold-country caterpillar hunter (*Calosoma frigidum*), USNM ENT 02024021
12. Cicada killer (*Sphecius speciosus*), USNM ENT 02024098
13. Virginia creeper sphinx (*Darapsa myron*), USNM ENT 02024018
14. Mydas fly (*Mydas clavatus*), USNM ENT 00891515
15. Regal moth (*Citheronia regalis*), USNM ENT 01978164
16. Swamp milkweed leaf beetle (*Labidomera clivicollis*), USNM ENT 02024030
17. Flea beetle (*Longitarsus cohongorooto*), USNM ENT 02024032
18. Spined micrathena (*Micrathena gracilis*), USNM ENT 01978558
19. Red milkweed beetle (*Tetraopes tetrophthalmus*), USNM ENT 02024028
20. Fall field cricket (*Gryllus pennsylvanicus*), USNM ENT 02024019
21. Eastern amberwing (*Perithemis tenera*), USNM ENT 00486341
22. Painted lady (*Vanessa cardui*), USNM ENT 02024075
23. Banded sphinx (*Eumorpha fasciatus*), USNM ENT 01978260

IMAGE CREDITS

6: Brittany M. Hance and James Di Loreto, Smithsonian Institution. **10–11:** James Di Loreto, Smithsonian Institution. **12:** Phillip R. Lee, Smithsonian Institution. **14:** James Di Loreto, Smithsonian Institution. **16:** James Di Loreto, Smithsonian Institution. **18:** Chip Clark, Smithsonian Institution. **19:** Chip Clark, Smithsonian Institution. **20:** Courtesy of Igor Krupnik, Smithsonian Institution. **20:** Courtesy of Gwyneira Isaac, Smithsonian Institution. **20:** Dena'ina Athabascan artist Joel Isaak shows Anchorage school students and chaperones how to soften moosehide with a stone scraper. Photo by Wayde Carroll, courtesy of the Smithsonian Arctic Studies Center, Alaska office. **32:** James Di Loreto, Smithsonian Institution. **35:** James Di Loreto and Tonda Phalen, Smithsonian Institution. **36:** Digitization Program Office, Smithsonian Institution. **36:** Catalog numbers E74339 and E74441, Department of Anthropology, Smithsonian Institution. Photo by James D. Tiller and James Di Loreto. **38:** Catalog number E394454-0, Department of Anthropology, Smithsonian Institution. Photo by James D. Tiller and Tonda Phalen. **40–41:** Catalog numbers E434512-0 and E151409-0, Department of Anthropology, Smithsonian Institution. Catalog number E434512-0 by Su'a Tupuloa Uilisone Fitiao, 2014. Photo by James D. Tiller. **43:** James D. Tiller and Fred Cochard, Smithsonian Institution. **44:** Phillip R. Lee, Smithsonian Institution. **47:** James Di Loreto, Smithsonian Institution. **49:** James Di Loreto, Smithsonian Institution. **50:** James Di Loreto, Smithsonian Institution. **52:** James D. Tiller and Fred Cochard, Smithsonian Institution. **54:** Brittany M. Hance, Smithsonian Institution. **57:** James D. Tiller, Smithsonian Institution. **59:** Brittany M. Hance, Smithsonian Institution. **60:** James Di Loreto, Smithsonian Institution. **62:** James Di Loreto, Smithsonian Institution. **65:** Phillip R. Lee, Smithsonian Institution. **67:** James Di Loreto, Smithsonian Institution. **68–69:** James Di Loreto, Smithsonian Institution. **70:** James Di Loreto, Smithsonian Institution. **73:** Phillip R. Lee and Fred Cochard, Smithsonian Institution. **75:** James Di Loreto, Smithsonian Institution. **76:** Catalog numbers E428722-0, E428718-0, E428727-0, and E428716-0, Department of Anthropology, Smithsonian Institution. Blacksmithing by Joaquin Flores Lujan. Photo by James D. Tiller and Tonda Phalen. **78:** Catalog number E430782-0, Department of Anthropology, Smithsonian Institution. Photo by James Di Loreto and Tonda Phalen. **81:** James Di Loreto, Smithsonian Institution. **82:** James Di Loreto, Smithsonian Institution. **84:** James Di Loreto, Smithsonian Institution. **85:** James Di Loreto and Tonda Phalen, Smithsonian Institution. **86:** James Di Loreto, Smithsonian Institution. **88–89:** James D. Tiller and Fred Cochard, Smithsonian Institution. **91:** James Di Loreto, Smithsonian Institution. **92:** Brittany M. Hance, Smithsonian Institution. **92:** Brittany M. Hance, Smithsonian Institution. **94:** James Di Loreto, Smithsonian Institution. **97:** James Di Loreto, Smithsonian Institution. **98–99:** James Di Loreto, Smithsonian Institution. **101:** James Di Loreto, Smithsonian Institution. **102:** James Di Loreto, Smithsonian Institution. **104:** James Di Loreto, Smithsonian Institution. **106–107:** James Di Loreto, Smithsonian Institution. **109:** James Di Loreto, Smithsonian Institution. **111:** Phillip R. Lee, Smithsonian Institution. **112:** Phillip R. Lee, Smithsonian Institution. **114:** Phillip R. Lee, Smithsonian Institution. **117:** Brittany M. Hance, Smithsonian Institution. **119:** James Di Loreto, Smithsonian Institution. **120:** Phillip R. Lee, Smithsonian Institution. **122:** Catalog numbers E380093-0 and E380094-0, Department of Anthropology, Smithsonian Institution. Photo by James D. Tiller. **123:** National Anthropological Archives, Smithsonian Institution, NAA INV 01778400. **124:** James D. Tiller, Smithsonian Institution. **124:** James Di Loreto, Smithsonian Institution. **127:** Brittany M. Hance and Phillip R. Lee, Smithsonian Institution. **129:** James D. Tiller, Smithsonian Institution. **130:** James Di Loreto, Smithsonian Institution. **133:** Brittany M. Hance, Smithsonian Institution. **135:** Brittany M. Hance, Smithsonian Institution. **136:** Phillip R. Lee, Smithsonian Institution. **138:** James Di Loreto, Smithsonian Institution. **141:** James D. Tiller, Smithsonian Institution. **142:** James D. Tiller and Phillip R. Lee, Smithsonian Institution. **145:** James D. Tiller, Smithsonian Institution. **146:** Phillip R. Lee, Smithsonian Institution. **148:** James Di Loreto, Smithsonian Institution. **151:** Phillip R. Lee and Fred Cochard, Smithsonian Institution. **152:** James D. Tiller, Smithsonian Institution. **155:** Phillip R. Lee, Smithsonian Institution. **156:** Phillip R. Lee, Smithsonian Institution.

158: Donald E. Hurlbert, Smithsonian Institution. 161: Phillip R. Lee, Smithsonian Institution. 163: James D. Tiller, Smithsonian Institution. 164: James Di Loreto, Smithsonian Institution. 166: Phillip R. Lee, Smithsonian Institution. 168: Phillip R. Lee, Smithsonian Institution. 170: Phillip R. Lee, Smithsonian Institution. 173: Catalog numbers E201483A-0 and E395238-0, Department of Anthropology, Smithsonian Institution. Photo by James Di Loreto and Tonda Phalen. 174–175: James Di Loreto, Smithsonian Institution. 177: James Di Loreto, Smithsonian Institution. 178: Phillip R. Lee, Smithsonian Institution. 181: James Di Loreto, Smithsonian Institution. 183: James Di Loreto, Smithsonian Institution. 184: Phillip R. Lee, Smithsonian Institution. 187: James Di Loreto, Smithsonian Institution. 189: Brittany M. Hance, Smithsonian Institution. 190: James D. Tiller, Smithsonian Institution. 193: James Di Loreto, Smithsonian Institution. 194: Phillip R. Lee, Smithsonian Institution. 197: Phillip R. Lee, Smithsonian Institution. 199: Phillip R. Lee, Smithsonian Institution. 200: James Di Loreto, Smithsonian Institution. 202: Phillip R. Lee and Fred Cochard, Smithsonian Institution. 205: James D. Tiller and Fred Cochard, Smithsonian Institution. 207: James Di Loreto, Smithsonian Institution. 208–209: Catalog number E307246-0, Department of Anthropology, Smithsonian Institution. Photo by Phillip R. Lee and James D. Tiller. 211: James Di Loreto, Smithsonian Institution. 212: Phillip R. Lee, Smithsonian Institution. 215: James D. Tiller and Phillip R. Lee, Smithsonian Institution.